STUDENT UNIT GUIDE

NEW EDITION

AQA A2 Geography Unit 3
Contemporary Geographical Issues

Amanda Barker, David Redfern
and Malcolm Skinner

Philip Allan Updates, an imprint of Hodder Education, an Hachette UK company, Market Place, Deddington, Oxfordshire OX15 0SE

Orders
Bookpoint Ltd, 130 Milton Park, Abingdon, Oxfordshire OX14 4SB
tel: 01235 827827
fax: 01235 400401
e-mail: education@bookpoint.co.uk
Lines are open 9.00 a.m.–5.00 p.m., Monday to Saturday, with a 24-hour message answering service.
You can also order through the Philip Allan Updates website: www.philipallan.co.uk

ISBN 978-1-4441-4779-7

First printed 2012
Impression number 5 4 3 2 1
Year 2016 2015 2014 2013 2012

Printed in Dubai

Hachette UK's policy is to use papers that are natural, renewable and recyclable products and made from wood grown in sustainable forests. The logging and manufacturing processes are expected to conform to the environmental regulations of the country of origin.

P01986

Contents

Getting the most from this book 4
About this book 5

Content guidance

Plate tectonics and associated hazards 8
Weather and climate and associated hazards 19
Ecosystems: change and challenge 30
World cities 39
Development and globalisation 49
Contemporary conflicts and challenges 62

Questions & Answers

Plate tectonics and associated hazards 74
Weather and climate and associated hazards 81
Ecosystems: change and challenge 86
World cities 92
Development and globalisation 100
Contemporary conflicts and challenges 107

Knowledge check answers 114
Index 118

Getting the most from this book

Examiner tips

Advice from the examiner on key points in the text to help you learn and recall unit content, avoid pitfalls, and polish your exam technique in order to boost your grade.

Knowledge check

Rapid-fire questions throughout the Content Guidance section to check your understanding.

Knowledge check answers

1 Turn to the back of the book for the Knowledge check answers.

Summary

Summaries

- Each core topic is rounded off by a bullet-list summary for quick-check reference of what you need to know.

Questions & Answers

Exam-style questions

Examiner comments on the questions
Tips on what you need to do to gain full marks, indicated by the icon ⓔ.

Sample student answers
Practise the questions, then look at the student answers that follow each set of questions.

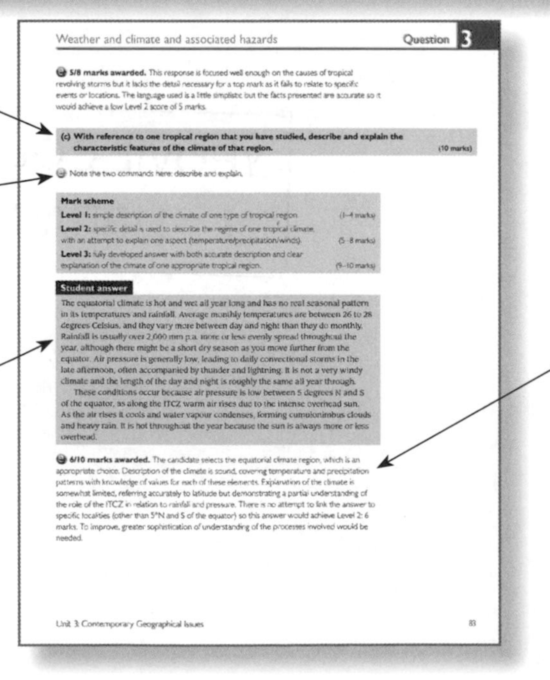

Weather and climate and associated hazards — Question 3

ⓔ **5/8 marks awarded.** This response is focused well enough on the causes of tropical revolving storms but it lacks the detail necessary for a top mark as it fails to relate to specific events or locations. The language used is a little simplistic but the facts presented are accurate so it would achieve a low Level 2 score of 5 marks.

(c) With reference to one tropical region that you have studied, describe and explain the characteristic features of the climate of that region. (10 marks)

ⓔ Note the two commands here: describe and explain.

Mark scheme

Level 1: simple description of the climate of one type of tropical region. (1–4 marks)

Level 2: specific detail is used to describe the regime of one tropical climate, with an attempt to explain one aspect (temperature/precipitation/winds). (5–8 marks)

Level 3: fully developed answer with both accurate description and clear explanation of the climate of one appropriate tropical region. (9–10 marks)

Student answer

The equatorial climate is hot and wet all year long and has no real seasonal pattern in its temperatures and rainfall. Average monthly temperatures are between 26 to 28 degrees Celsius, and they vary more between day and night than they do monthly. Rainfall is usually over 2,000 mm p.a. more or less evenly spread throughout the year, although there might be a short dry season as you move further from the equator. Air pressure is generally low, leading to daily convectional storms in the late afternoon, often accompanied by thunder and lightning. It is not a very windy climate and the length of the day and night is roughly the same all year through.

These conditions occur because air pressure is low between 5 degrees N and S of the equator, as along the ITCZ warm air rises due to the intense overhead sun. As the air rises it cools and water vapour condenses, forming cumulonimbus clouds and heavy rain. It is hot throughout the year because the sun is always more or less overhead.

ⓔ **6/10 marks awarded.** The candidate selects the equatorial climate region, which is an appropriate choice. Description of the climate is sound, covering temperature and precipitation patterns with knowledge of values for each of these elements. Explanation of the climate is somewhat limited, referring accurately to latitude but demonstrating a partial understanding of the role of the ITCZ in relation to rainfall and pressure. There is no attempt to link the answer to specific localities (other than 5°N and S of the equator) so this answer would achieve Level 2: 6 marks. To improve, greater sophistication of understanding of the processes involved would be needed.

Unit 3: Contemporary Geographical Issues 83

Examiner commentary on sample student answers
Find out how many marks each answer would be awarded in the exam and then read the examiner comments (preceded by the icon ⓔ) following each student answer.

About this book

All students of A2 geography who are following the AQA specification have to study Unit 3: Contemporary geographical issues. You need to understand:

- the key ideas of the content of the unit
- the nature of the assessment material by looking at sample structured and essay questions
- how to perform well in examinations

This guide provides information to help you satisfy each of the above requirements.

The content guidance section summarises the essential information of Unit 3. It will make you aware of the material to be covered and learnt.

The question and answer section includes sample questions in the style of the examinations for each of the physical and human options. Example student responses are given at a range of levels. Each answer is followed by a detailed examiner's commentary. It is suggested that you read through the relevant topic area in the content guidance section before attempting a question from the question and answer section, and read the specimen answers only after you have tackled the question yourself.

Unit 3 Scheme of assessment

The exam lasts 2½ hours, and consists of structured questions and essay questions carrying 90 marks in total. Students must answer three questions:

- **Section A** One structured question (25 marks) from one of the three physical options (Plate tectonics and associated hazards, Weather and climate and associated hazards, Ecosystems: change and challenge).
- **Section B** One structured question (25 marks) from one of the three human options (World cities, Development and globalisation, Contemporary conflicts and challenges).
- **Section C** One essay question (40 marks) from an option not answered above.

Therefore all students must study *at least three* options: one physical, one human and one other (either physical or human).

Stretch and challenge

Assessments at A2 provide greater stretch and challenge for all candidates. This includes the use of more open-ended questions that require the responses to be structured by the candidates. This means:

- the use of a variety of demanding command words in questions — for example 'comment on', 'discuss', 'analyse', 'evaluate'
- greater connectivity between sections of questions
- the use of a wider range of question types to address different skills (for example open-ended questions) and a requirement for the detailed use of a range of case studies.

For example, within Unit 3, the command words 'discuss' and 'evaluate' can be used in Sections A and B, and in Section C the further demanding stems of 'critically evaluate', 'analyse' and 'assess' may be used. The use of open-ended questions such as 'To what extent do you agree...' provides opportunities for thorough, well-developed and critical responses. The command 'comment on' is often used in data-stimulus questions. This requires you to examine the stimulus material provided and then make statements arising from the material that are relevant, appropriate and geographical, but not directly evident.

Synoptic assessment

The definition of synoptic assessment in the context of geography is as follows:

> Synoptic assessment involves assessment of candidates' ability to draw on their understanding of the connections between different aspects of the subject represented in the specification and demonstrate their ability to 'think like a geographer'.

Synoptic assessment is included in the essay section (Section C) of Unit 3. The mark scheme for the essay questions is given below.

Mark scheme for the essay questions

Assessment criteria	**Level 1 1–10 marks (midpoint 6)**	**Level 2 11–20 marks (midpoint 16)**	**Level 3 21–30 marks (midpoint 26)**	**Level 4 31–40 marks (midpoint 36)**
Knowledge of content, ideas and concepts	Basic grasp of concepts and ideas; points lack development or depth	The answer is relevant and accurate. Reasonable knowledge. Imbalanced theories	Sound and frequent evidence of thorough, detailed and accurate knowledge	Strong evidence of thorough, detailed and accurate knowledge
Critical understanding of the above	Incomplete, basic	Reasonable critical understanding of concepts and principles with some use of specialist vocabulary	Sound and frequent evidence of critical understanding of concepts and principles, and of specialist vocabulary	Strong evidence of critical understanding of concepts and principles and of specialist vocabulary
Use of examples/ case studies to support argument	Superficial	Examples show imbalances and/or lack detail	Examples are developed, balanced and support the argument	Examples are well developed and integrated
Maps/Diagrams	None	Ineffective	Effective	Fully integrated
Evidence of synopticity	No evidence	Limited	Strong	Full
Connections between different aspects of the subject		Some ability to identify, interpret and synthesise some of the material	Some ability to identify, interpret and synthesise a range of material	There is a high level of insight, and an ability to identify, interpret and synthesise a wide range of material with creativity
'Thinking like a geographer'		Limited ability to understand the roles of values, attitudes and decision-making processes	Some ability to understand the roles of values, attitudes and decision-making processes	Evidence of maturity in understanding the role of values, attitudes and decision-making processes

Assessment criteria	Level 1 1–10 marks (midpoint 6)	Level 2 11–20 marks (midpoint 16)	Level 3 21–30 marks (midpoint 26)	Level 4 31–40 marks (midpoint 36)
Quality of argument — the degree to which an argument is constructed, developed and concluded	Language is basic; arguments are partial, over simplified and lacking clarity. Little or no sense of focus of task	Arguments are not fully developed or expressed clearly, and the organisation of ideas is simple and shows imbalances. Some sense of focus of task	Explanations, arguments and assessments or evaluations are accurate, direct, logical, purposeful, expressed with clarity and generally balanced. Clear sense of focus of task	Explanations, arguments and assessments or evaluations are direct, focused, logical, perceptive, mature, purposeful, and are expressed coherently and confidently, and show both balance and flair

In deciding the overall level, each of the assessment criteria is awarded a level, and then a best fit is put into practice.

Content guidance

This section gives an overview of the key terms and concepts covered in Unit 3: Contemporary geographical issues.

There is detailed guidance on the content on each of the options:
- Plate tectonics and associated hazards
- Weather and climate and associated hazards
- Ecosystems: change and challenge
- World cities
- Development and globalisation
- Contemporary conflicts and challenges

For each option advice is also given on the nature of case study support that is necessary.

Plate tectonics and associated hazards

The theory of plate tectonics

Plate tectonics is the name given to a set of concepts or theories that try to explain the formation and distribution of the Earth's major structural features by reference to the relative movement of the plates that make up the Earth's surface. These structural features include the continents, ocean basins, mountain ranges, oceanic trenches and volcanoes.

The structure of the Earth

Before the development of the theory, Earth scientists divided the interior of the Earth into three layers:
- **Core** — made up of dense rocks containing iron and nickel alloys and divided into a solid inner core and a molten outer area with a temperature of over 5,000°C.
- **Mantle** — made up of molten and semi-molten rocks containing lighter elements such as silicon and oxygen.
- **Crust** — of even lower density because more of the lighter elements are present, the most abundant being silicon, oxygen, aluminium, potassium and sodium. The crust varies in thickness from 6–10 km beneath the oceans to 30–40 km below the continents. The crust is thickest under the highest mountains, up to 70 km.

More recent research has retained this simple threefold division but has divided the **crust** and **upper mantle** into two layers:
- **Lithosphere** — the crust and the rigid upper section of the mantle, approximately 80–90 km thick, and divided into a number of plates.

- **Asthenosphere** — semi-molten material below the lithosphere on which the plates of the lithosphere 'float'.

The basis of plate tectonic theory

Plate tectonic theory is based upon the differences of the properties of the lithosphere and the asthenosphere:

- Hot spots around the core generate thermal convection currents within the asthenosphere.
- Magma rises towards the surface and spreads before cooling and sinking.
- This circulation of magma is the vehicle upon which the crustal plates of the lithosphere move.
- This continuous process forms new land along the boundaries where plates are moving apart (constructive boundaries) and destroys older crust where plates are moving together (destructive boundaries).

Knowledge check 1

What is the evidence that the circulation of magma is a continuous process?

Table 1

Continental crust	Oceanic crust
30–70 km thick	6–10 km thick
Over 1,500 million years old	Less than 200 million years old
Lighter density (2.6)	More dense (3.0)
Variety of rocks, mainly granite	Mainly basaltic rocks
Main minerals: silicon, aluminium and oxygen (known as SIAL)	Main minerals: silicon, magnesium and oxygen (known as SIMA)

Evidence for plate tectonic theory

Alfred Wegener published his theory of continental drift in 1912. He maintained that about 300 million years ago a single continent existed which he named Pangaea. This later split into Laurasia in the north and Gondwanaland in the south. Today's continents are the result of further splitting.

The evidence that Wegener claimed supported his theory was:

- The 'fit' of the coastlines of Africa and South America.
- Evidence for a late-Carboniferous glaciation exists in deposits in India, South America and Antarctica. These deposits must have been formed together and subsequently moved.
- Rock sequences in northern Scotland agree closely with those found in eastern Canada, indicating that they were laid down in the same position.
- Fossil brachiopods found in some Indian limestones are comparable with similar fossils found in Australia.
- Fossil remains of the reptile Mesosaurus are found in both southern Africa and South America. It is unlikely that it could have developed in two such different locations.
- The fossilised remains of a plant that existed when coal was being formed are found in both India and Antarctica.

Examiner tip

You should know the evidence that was used to support plate tectonic theory when it was first proposed and the modern evidence that has led to the acceptance of the theory.

Wegener could not explain the movement of the continents and so his ideas gained little ground. In the second half of the twentieth century the theory came to be accepted because of:

- The discovery of mid-oceanic ridges.

- Palaeomagnetism — evidence of sea-floor spreading is gained from an examination of the polarity of the rocks that make up the ocean floor. Iron particles in lava are aligned with the Earth's magnetic field. At regular intervals the polarity of the Earth reverses; this results in a series of magnetic stripes with the sea-floor rocks aligned alternately towards north and south poles. This striped pattern, which is mirrored exactly on either side of a mid-oceanic ridge, suggests that the ocean crust is slowly spreading away from the boundary. Also, the ocean crust gets older with distance from the mid-ocean ridge.
- The discovery of ocean trenches where large areas of ocean floor are pulled downwards and destroyed. If plates are being built up and the Earth is getting no bigger, then crust has to be destroyed somewhere.

Knowledge check 2

Explain the process of sea-floor spreading.

Features of plate margins

Constructive (divergent) margins

This is where plates are moving away from each other and new crust is continually being created (particularly under the oceans). The main features formed are:

- **Mid-oceanic ridges** — these are long (they run for thousands of kilometres), high (rising in some areas over 3,000 m from the ocean floor) and often with complex structures of rifts and scarps. Transform faults occur at right angles to the main plate boundary. Shallow-focus earthquakes also occur.
- **Volcanoes** — volcanic activity occurs along mid-oceanic ridges, sometimes rising above sea level to produce islands (e.g. Surtsey, off Iceland). Such volcanoes are formed from basaltic lava which has a low viscosity and flows great distances, creating volcanoes with gentle sides. Volcanoes also form in association with rift valleys, especially in east Africa. These volcanoes are very different from those associated with mid-oceanic ridges.
- **Rift valleys** — form at constructive boundaries on continental areas due to the fracturing of brittle crust. Areas of crust drop down between parallel faults to form the feature. The best known is the east African rift valley which extends from Mozambique through eastern Africa and the Red Sea to Jordan in the middle east, a distance of 5,500 km. In some areas the inward-facing scarps are over 600 m above the valley floor. This African rift is thought to be an emerging plate boundary as east Africa splits from the rest of the continent.

Examiner tip

You must know the detailed differences between constructive and destructive margins.

Destructive (convergent) margins

With two types of plates (oceanic/continental), there are three types of convergence:

- Oceanic and continental, e.g. off western South America where the denser oceanic Nazca plate is moving under (known as subduction) the lighter continental South American plate.
- Oceanic plate meeting oceanic plate, e.g. in the western Pacific Ocean where the Pacific plate is moving under the smaller Philippine plate.
- Two continental plates, e.g. the Indo-Australian plate meeting the Eurasian plate in southern Asia. Here the two plates have a lower density than underlying layers. Consequently there is little subduction and the plate edges are forced upwards into fold mountains.

Earthquakes are associated with all three types. Shallow, intermediate and deep earthquakes are associated with oceanic/continental and oceanic/oceanic convergences, but only shallow earthquakes are found in continental/continental collisions.

The main features formed when plates converge are:

- **Ocean trenches** — as the denser plate is subducted, the ocean floor is pulled down to form a trench. Examples include the Peru-Chile trench off western South America and the Mariana trench in the western Pacific (one of the deepest points of the world's oceans).
- **Fold mountains** — when oceanic and continental plates meet, sediments accumulating on the continental shelf are forced upward and are deformed by folding and faulting, e.g. Andes (western South America). When continental plates meet, the edges are forced up, e.g. the Himalayas, the highest point of the planet.
- **Volcanoes** — when a plate is subducted, the deeper it is pushed the hotter its surroundings. This heat, and the heat generated by friction, cause the plate to melt in an area known as the Benioff zone. This molten material is lighter than the surrounding asthenosphere and rises towards the surface as plutons of magma. When it reaches the surface, volcanoes are formed. This is andesitic lava, which is viscous (doesn't flow easily) and forms composite and explosive volcanoes.
- **Island arcs** — the magma comes to the surface under water to form a line of volcanoes, e.g. the Mariana islands formed in association with the Mariana trench. The islands of the West Indies are another good example.

Knowledge check 3

Explain in detail the process of subduction. How can subduction affect the surface of the Earth?

Conservative margins

Here, plates slide past each other and there is no destruction or construction of crust. There is also no volcanic activity. However, movements of this kind create stresses between the plate edges that, when released, cause shallow-focus earthquakes. The best known example is in western North America where the Pacific and North American plates are sliding past each other at different rates, forming the San Andreas fault zone in California.

Hot spots

Volcanic activity can also occur away from plate boundaries. A hot spot is where a concentration of radioactive elements in the mantle causes a plume of magma to rise towards the surface, eating into the oceanic plate and eventually forming volcanoes on the surface. The Hawaiian islands, in the middle of the Pacific Ocean, are believed to be situated over a hot spot. They also show the movement of the Pacific plate: as the hot spot is stationary, the line of islands demonstrates that the plate has moved over it in recent geological time. At one end of the chain the volcanic islands are now extinct and worn away, at the other the volcanic activity is almost constant (Hawaii).

Knowledge check 4

Explain why hot spots can demonstrate that plates are mobile.

Summary of the relationship of tectonic activity to plate margins

Table 2 Relationship of tectonic activity to plate margins

Plate margin	Movement of plates	Tectonic features	Examples
Constructive	Divergent: two plates moving away from each other	New crust is formed from upwelling magma: mid-oceanic ridges, effusive ridge (shield) volcanoes, shallow-focus earthquakes, median rift valleys	Mid-Atlantic ridge
		Continental rift valleys	East African rift valley
Destructive (1) Subduction	Convergent: two plates moving towards each other	(1a) Oceanic to oceanic: trenches, island arcs, explosive volcanoes, earthquakes (shallow, intermediate and deep)	On the margins of Pacific plate, with subduction under other, separate sections of the plate — Tonga trench
		(1b) Oceanic to continental: trenches, fold mountains, explosive volcanoes, earthquakes (shallow, intermediate and deep)	Andean type: Nazca plate subducting under South American plate
(2) Collision		(2) Continental to continental: fold mountains, shallow-focus earthquakes	Himalayan type: Indian plate colliding with Eurasian plate
Conservative	Two plates shearing past each other	Shallow-focus earthquakes	San Andreas fault: Pacific plate and North American plate
Not at plate boundaries	Hot spots: may be near the centre of a plate	Plume volcanoes	Hawaiian islands: Emperor seamount chain

Examiner tip
Make sure for the examination that all the examples given in Table 2 are known to you. This is particularly important with regard to location.

Vulcanicity

The distribution of volcanic activity

Volcanic activity is associated with the following:

- ocean ridges — where plates are moving apart
- rift valleys — where parts of plates are moving apart
- subduction zones — where one plate is destroyed at depth causing magma to rise to the surface
- hot spots — where plates move over rising plumes of magma

Knowledge check 5
Can you locate these areas of volcanic activity on a world map?

Volcanic eruptions

When magma is forced to the surface, only a small amount reaches it. Most solidifies in the crust, producing a range of features that are exposed by later erosion:

- **Batholith** — large mass, often dome-shaped and consisting of granite, e.g. Dartmoor. The rock around is altered by heat and pressure to form a **metamorphic aureole**.
- **Laccolith** — smaller than a batholith, tends to be lens shaped, e.g. Eildon Hills, Scotland.
- **Dykes** — vertical intrusions, cutting across bedding planes, e.g. on the Isle of Mull.
- **Sills** — horizontal intrusions along bedding planes, e.g. Great Whin Sill (northern England).

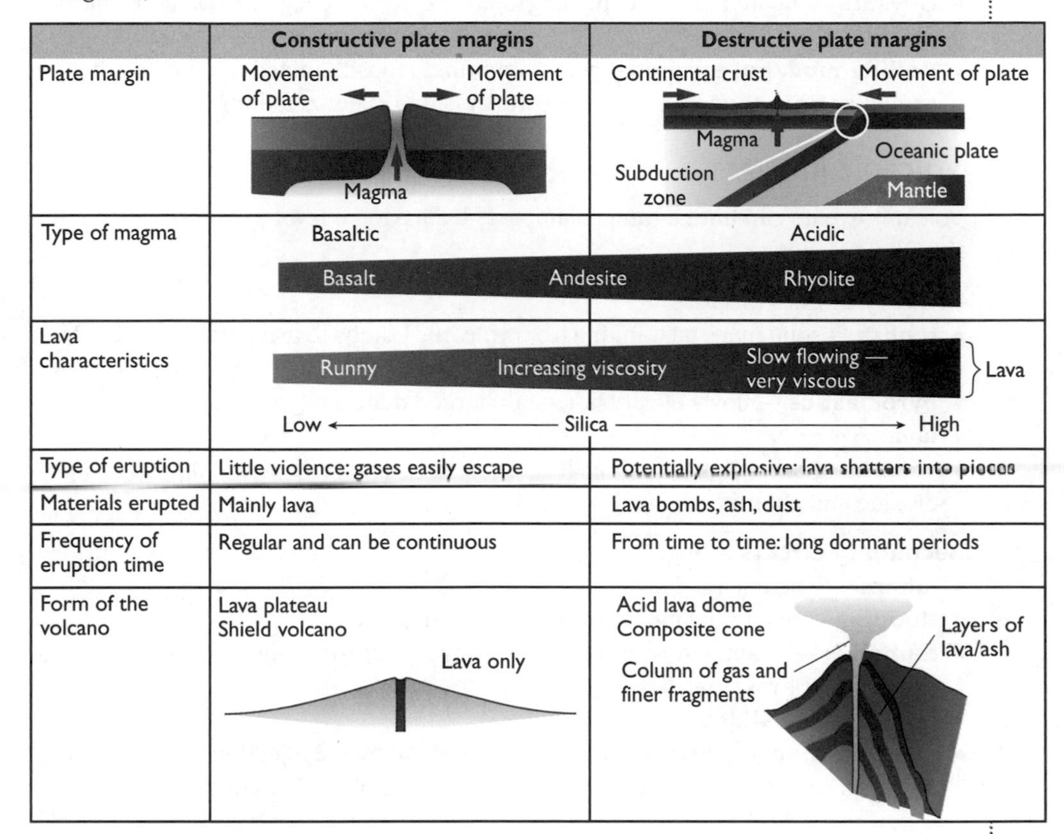

	Constructive plate margins	Destructive plate margins
Plate margin	Movement of plate / Movement of plate / Magma	Continental crust / Movement of plate / Magma / Oceanic plate / Subduction zone / Mantle
Type of magma	Basaltic — Basalt, Andesite	Acidic — Rhyolite
Lava characteristics	Runny — Increasing viscosity	Slow flowing — very viscous } Lava
	Low ← Silica	→ High
Type of eruption	Little violence: gases easily escape	Potentially explosive: lava shatters into pieces
Materials erupted	Mainly lava	Lava bombs, ash, dust
Frequency of eruption time	Regular and can be continuous	From time to time: long dormant periods
Form of the volcano	Lava plateau Shield volcano Lava only	Acid lava dome Composite cone Column of gas and finer fragments Layers of lava/ash

Figure 1 The effect of different plate margins on volcanic eruptions and landforms

Surface eruptions involve two forms of lava:

- **basaltic lava** — low in silica and therefore tends to be very fluid
- **andesitic lava** — silica-rich (acid) magma which is often viscous. Often solidifies before reaching the surface, causing pressure which leads to explosive activity

Surface features formed include:

- **Lava plateaux** — formed from fissure eruptions. Runny lava spreads great distances, e.g. Antrim (Northern Ireland).
- **Basic/shield volcanoes** — also formed from fluid lava which produces volcanoes with gentle sides but very wide bases, e.g. Hawaiian islands.
- **Acid/dome volcanoes** — steep-sided, produced from viscous lava (rhyolite), e.g. Puy region (central France).
- **Ash and cinder cones** — formed from ash, cinders and volcanic bombs ejected from a crater, e.g. Paricutin (Mexico).

- **Composite cones** — the classic pyramid-shape volcanoes consisting of layers of lava and ash, e.g. Mt Etna (Sicily).
- **Calderas** — formed when huge explosions remove the cone of a volcano, leaving an opening several kilometres across, e.g. Krakatoa (Indonesia).

Minor volcanic features include:
- **Solfatera** — small volcanic areas without cones, mainly consisting of gases escaping to the surface, e.g. around Bay of Naples (Italy).
- **Geysers** — heated water, exploding onto the surface, e.g. Yellowstone National Park (USA).
- **Boiling mud/hot springs** — water and fine deposits mixed and heated, but not exploding onto surface, e.g. Iceland, North Island (New Zealand).

Volcanic impacts

Volcanic events can have a range of impacts from a local level to events which affect the whole planet.

Examiner tip

It is important that you are clearly able to differentiate between primary and secondary effects. You should also be able to give examples of where such effects have impacted upon human populations.

Primary effects include:
- **tephra** — solid material, ranging from volcanic bombs to ash particles ejected into the atmosphere
- **pyroclastics** — flows of very hot, gas-charged material (gas and tephra)
- **lava**
- **volcanic gases** — carbon dioxide, carbon monoxide, hydrogen sulphide, sulphur dioxide and chlorine

Secondary effects include:
- **lahars** — volcanic mud flows
- **flooding** — due to the melting of ice caps and glaciers
- **tsunamis** — giant sea waves generated by caldera-forming events such as the explosion of Krakatoa
- **volcanic landslides**
- **climatic change (short term)** — brought about by the injection of vast amounts of debris into the atmosphere, which can reduce global temperatures

Seismicity

Causes of earthquakes

The movement of the Earth's crust means there is a slow build up of stress within the rocks. When this pressure is suddenly released, parts of the surface experience an intense shaking motion for a brief period (no more than a few seconds) — this is an earthquake.

Features of earthquakes include:
- **focus** — the point at which this pressure release occurs
- **epicentre** — the point on the surface immediately above the focus
- **seismic waves** — radiate from the focus, like ripples in water. **Primary waves (P)** travel fastest, **secondary waves (S)** travel at half that speed, and **surface waves (L)** are slowest. The study of seismic waves has allowed a picture of the interior of the Earth to be built up

Distribution of earthquakes

Earthquakes are mainly found at:

- **plate boundaries** — particularly at conservative margins where plates are moving alongside each other (e.g. California). Very destructive earthquakes also tend to occur where plates are moving towards each other (convergence)
- **old fault lines**

Magnitude and frequency

Magnitude is measured on two scales. The **Richter scale** is logarithmic with the energy released by the earthquake being proportional to the magnitude on the scale. The scale is from 1 (very slight, only detectable by seismographs) to 9/10 (very rare, ground seen to shake with massive damage over a wide area). The **Mercalli scale** measures the intensity of the event and its impact. This runs from 1 (1/2 on the Richter scale) to 12 (ground shaking, equivalent to 8.5 Richter). The **frequency** of events can be observed using seismic records, although these go back only to 1848 when the seismograph was first developed.

The effects of earthquakes

The initial effect is **ground shaking**. This is followed by a number of secondary effects:

- **soil liquefaction** — soil starts to behave as a fluid
- **landslides/avalanches** — slope failure resulting from ground shaking
- **collapsing buildings/transport systems**
- **destruction of service provision such as water, gas, electricity**
- **fires** — ruptured gas mains and fallen electricity pylons
- **flooding** — from dam bursts
- **disease and food shortages**
- **tsunamis** (see below)

Tsunamis

A tsunami (Japanese for 'harbour wave') is a giant sea wave generated by a shallow-focus underwater earthquake (there are other causes such as volcanic eruptions, large surface landslides and underwater debris slides). Tsunamis have been known to reach heights in excess of 25 m and the event can consist of a number of waves, the largest not necessarily being the first. Tsunami effects depend upon:

- height of the waves
- distance travelled
- length of the event
- possibility of issuing warnings
- coastal geography
- coastal land use and population density

Tsunamis wash boats and structures inland and the backwash may carry them back out to sea. The effects can be felt 500/600 m inland, depending upon the coastal geography. Ninety per cent of all tsunamis are generated within the Pacific Ocean basin and are associated with tectonic activity taking place around its edges.

Knowledge check 6

What is the relationship between the location of volcanoes and seismic activity?

Examiner tip

As with volcanic activity, it is important to differentiate between primary and secondary effects. Some authorities take a slightly different approach to what is primary and what is secondary in this context. It doesn't really matter what approach you use in an examination as long as your distinction is coherent and one that you are able to justify in your answer.

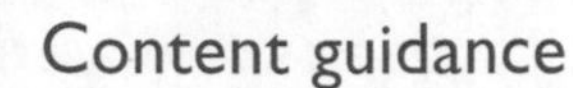

Knowledge check 7

What determines the impact of a tsunami?

Examiner tip

As far as is possible, you should always try to give examples/case studies that are up to date. At the time of writing, the Japanese tsunami of March 2011 was the most recent of its kind and it would certainly be to your advantage to have some knowledge of this event.

The most devastating tsunami of recent times was that of 26 December 2004, which affected many coastal areas around the Indian Ocean. It was caused by a powerful submarine earthquake (estimated at 9.0 on the modified Richter scale) off the northwestern tip of Sumatra. The main effects of the tsunami were:

- an estimated 300,000 people were killed
- tens of thousands were injured by the force of the waves and the debris they carried
- many European tourists were either killed or injured
- whole towns and villages were swept away
- millions of people were made homeless
- the tourist infrastructure around the Indian Ocean was badly damaged
- communications were badly affected, particularly where bridges, railway lines and roads were swept away
- the coastal economy of several nations was badly damaged, particularly the fishing industry with thousands of boats swept away or damaged
- medical facilities were damaged, resulting in medical aid having to be brought in from outside the region

As a result, a warning system has been set up in countries which surround the Indian Ocean basin.

Volcanic and seismic events

You are required to examine **two** case studies of recent events (ideally within the last 30 years) of each of the above. The two studies should be taken from contrasting areas of the world.

Volcanic events

There are a number of volcanic eruptions that have taken place in the last 30 years which are well documented and ideal for study. These include:

- Nevado del Ruiz (Colombia), 1985
- Pinatubo (Philippines), 1991
- Mt Etna (Sicily), 1991–93
- Soufrière Hills (Montserrat), 1995–97
- Nyiragongo (Congo), 2002
- Chaiten (Chile), 2008

It is difficult to compare the interaction of volcanic events and people. This is because in each case the type of volcanic activity is often very different and therefore has a different impact on people.

In the Nyiragongo event in 2002 the lava escaped quickly from the volcano and overwhelmed areas in and around Goma. At least one-third of the town was destroyed, along with the airport, the commercial centre and many of its medical facilities. A red alert was issued that enabled most people to escape from the lava, so the estimated death toll was low, at only 147. Vast numbers fled the area, however (estimated 350,000), and this put enormous pressure on neighbouring areas, particularly Gisenyi in Rwanda. Sulphurous lava polluted Lake Kivu, a major source of drinking water, and many people fell ill from the effects of the smoke and fumes, and from drinking the contaminated water. There was also much looting of abandoned properties. The United Nations organised relief for the area, although it was 2 days after the eruption before emergency rations began to arrive. The UN also set up camps to house displaced people.

The eruption on Mt Etna between 1991 and 1993 posed a very different threat and response. The lava flowed much more slowly than at Nyiragongo and originated at a height that did not immediately threaten people or property. The lava flows did eventually threaten some communities, particularly Zefferana, and the authorities acted to lessen the danger. They tried to stop or simply slow down the flow by creating earth barriers and eventually dropping concrete blocks into the lava tubes and then blasting openings in those tubes to divert the lava away. This was extremely successful and the lava did not reach the village before the eruption ended in early 1993. Attempts to control other lava flows on Mt Etna have not always been so successful. The eruption of 2002 destroyed a ski centre on the flanks of the volcano and large amounts of ash fell on the city of Catania.

Earthquake events

There are a number of well documented earthquakes that have occurred in the last 30 years. You could look at some of the following:

- Spitak (Armenia), 1988
- Northridge, Los Angeles (USA), 1994
- Kobe (Great Hanshin) (Japan), 1995
- Izmit (Turkey), 1999
- Gujurat (India), 2001
- Bam (Iran), 2003
- off northwest Sumatra (Indonesia), 2004
- Kashmir (Pakistan), 2005
- Sichuan (China), 2008
- L'Aquila (Italy), 2009
- Haiti (2010)
- Tohoku (Japan) (2011)

Examiner tip

For each case study, you should examine the nature of the volcanic/seismic event, its impact, the management of the hazard, and responses to the event.

A good comparison is the Gujurat earthquake of 2001 with the Los Angeles earthquake of 1994 (Table 3). Although the Gujurat event was slightly more powerful, the high death toll reflects the relative structural weakness of many of the buildings which simply collapsed upon their inhabitants. Many of the medical facilities and much of the basic infrastructure were destroyed in Gujurat. However, in Northridge, although 11 hospitals were damaged, it was possible to bring high-class medical facilities into the area. The difference between a relatively poor developing area and a developed country is clearly seen in the estimated cost of the damage: in India, it is comparatively low at $4–5 billion, whereas in California, the cost of replacing buildings, medical facilities and damaged infrastructure was estimated to be considerably higher at over $30 billion.

Examiner tip
Make sure that all your examples/case studies are as detailed as possible. It is also important to be as up to date as you can be. You should research new events as they occur.

Table 3 The Gujurat and Northridge earthquakes compared

	Gujurat	Northridge
Intensity	7.9 Richter	6.7 Richter
Death toll	20,000 (poss 30,000)	57
Seriously injured	160,000	1,500
Number homeless	1 million	20,000
Buildings affected	345,000 destroyed, 800,000 with some damage	12,500 (moderate/serious damage)
Other features	Large number of hospitals/clinics destroyed In small towns over 90% of dwellings destroyed (e.g. Buj) Some illnesses such as diarrhoea	11 hospitals damaged
Estimated cost	$4–5 billion	Over $30 billion

Hazard management

Studying two contrasting case studies of the hazards will give you some idea of how management attempts to cope with such events. It will not give you the whole picture. Some of the ways in which people seek to manage the effects of volcanic activity and earthquakes are as follows.

Volcanic activity

Prediction

- study the eruption history of the volcano
- measure gas emissions, land swelling, groundwater levels
- measure the shock waves generated by magma travelling upwards

Protection

- hazard assessment — trying to determine the areas of greatest risk which should influence land use planning
- dig trenches to divert the lava
- build barriers to slow down lava flows
- explosive activity to try to divert a lava flow
- pour water on the lava front to slow it down

Knowledge check 8
Why is it impossible to prevent volcanic and earthquake hazards? Because prevention is ruled out, how does this determine the way that people react to the threat of these hazards?

Earthquakes

Prediction

- study groundwater levels, release of radon gas and animal behaviour
- monitor fault lines and local magnetic fields
- study fault lines to look for 'seismic gaps' where the next earthquake may occur

Prevention (impossible?)

- keep the plates sliding past each other rather than 'sticking' and then releasing; suggestions include using water and oil

Protection

- build hazard-resistant structures: install a large weight that can move with the aid of a computer program to counteract stress; large rubber shock absorbers in foundations; cross-bracing to hold the building when it shakes
- retro-fit older buildings and elevated motorways with such devices
- educate people in survival strategies and encourage earthquake drills
- advise people in assembling earthquake kits, to include stored water, canned food, clothing/bedding, first aid kit, torch, batteries, can opener, matches, toilet paper and a small fire extinguisher
- install 'smart' meters that cut off gas supplies at a certain tremor threshold
- keep emergency services well organised with the correct gear (heavy lifting)
- plan land use to avoid certain buildings being constructed in high risk areas

Examiner tip

Be able to give actual examples of where such methods have been employed. You should also be able to evaluate the levels of success that have been achieved.

Summary

Plate tectonics and associated hazards

- Plate tectonics refers to a set of concepts and theories that try to explain the formation and distribution of the Earth's major structural features. It involves the layers of the Earth, plates, and plate movements through convection currents.
- These ideas were brought together in Alfred Wegener's theories which, although originally rejected, now explain the distribution of the world's major surface features.
- The Earth's surface is divided into a number of plates, which interact. This interaction brings about many of the surface features such as mid-oceanic ridges, rift valleys, oceanic trenches, fold mountains and island arcs, and is responsible for widespread volcanic activity and earthquakes.
- Volcanic activity can occur away from plate boundaries where hot spots are found over a plume of magma that is rising to the surface.
- Volcanic activity takes place in a number of different types of location and is responsible for producing surface features such as volcanoes, lava plateaux and calderas, with a number of minor features (solfatera, geysers and hot springs). Some rising magma solidifies in the crust, producing batholiths, laccoliths, dykes and sills.
- Earthquakes occur when the build up of stress within rocks is suddenly released, leading to a period of intense shaking. There are a number of effects that result from such action, which can be divided into primary and secondary categories.
- You need to provide two contrasting case studies of a volcanic event and two of an earthquake. Each study should include the nature of the event, its impact and the management of, and responses to, the event. To get a contrast, one event should be in a developing country and the other from a developed area. It is also very important to understand other aspects of management that are not covered by the case studies chosen.

Weather and climate and associated hazards

Major climate controls

The structure of the atmosphere

The Earth's layered atmosphere officially extends to 1,000 km from the surface, but 99% of the gases necessary for plant and animal life lie much closer — within 40 km.

Examiner tip
Ensure that you can draw an accurate diagram to show the layers within the Earth's atmosphere.

The gases that form the atmosphere are held to the Earth by gravity and include nitrogen (78.09%), oxygen (20.95%) and other gases in smaller amounts, such as argon, carbon dioxide, helium, methane and radon. Most of our climate and weather processes operate in the layer of the atmosphere closest to the ground called the troposphere, which is 16–17 km deep. The top of this layer, the tropopause, acts as a ceiling to our weather systems. The stratosphere lies above the tropopause and this is the layer of the atmosphere containing the greatest amount of ozone, which essentially shields the Earth from most of the sun's harmful ultraviolet radiation. The mesosphere and the thermosphere are the two highest layers. Although air pressure decreases with altitude, temperatures actually increase with distance in the outermost thermosphere layer.

The atmospheric heat budget

This is the balance between the incoming solar radiation (insolation) and outgoing radiation from the planet. Overall, the Earth's surface has a net gain of energy in the tropics and a net loss in the polar regions. However, as the tropics are not getting hotter nor the poles colder, there is obviously a transfer of energy between these regions.

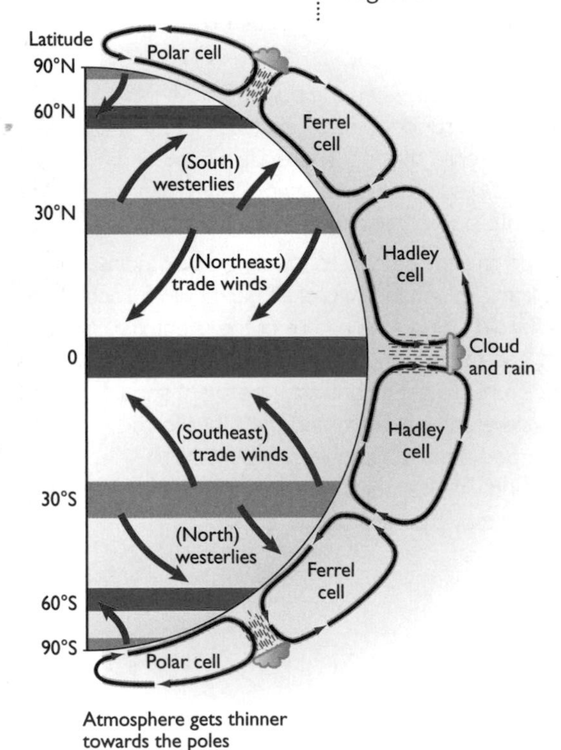

Figure 2 The tri-cellular model of atmospheric circulation

The general atmospheric circulation

This is the generalised pattern of wind and pressure belts within the Earth's atmosphere. Although complex in reality, certain movements occur regularly enough for us to recognise regions of low and high air pressure and general wind directions (Figure 2).

There is a net surplus of energy in the lower latitudes close to the equator and a net deficit close to the poles. There is a continuous transfer of heat from the tropics towards the poles by winds and ocean currents. Such transfers have been explained using global circulation models or theories, such as the tri-cellular model. This model, although crucial to our understanding of global circulation, does not allow for the influence of high-level jet streams in the redistribution of energy.

Planetary surface winds

Winds result from differences in air pressure and always move from areas of high to low pressure on the Earth's surface. The tri-cellular model helps to explain the global pattern of surface winds but, on a smaller scale, winds can also be influenced by the presence of land and sea. The British Isles lie in the path of the southwesterlies which generally bring on-shore winds to the west coast. However, off-shore breezes can occur during the night when the sea is relatively warm, creating local air pressure variations between land and sea.

Latitude

Latitude has a significant influence on global temperature patterns. Insolation or solar heating varies globally for two reasons:

- There is marked variation in the length of daylight with distance from the equator. During the winter months in the high latitudes there is very little opportunity for insolation as the days are very short.
- The angle of incidence of the sun is higher in the tropics than it is at the polar latitudes. The sun's rays are therefore more concentrated because they travel a shorter distance through the atmosphere before hitting the Earth. They also heat up a much smaller surface area in the low latitudes, closer to the equator.

Examiner tip

A common exam howler is to mix up latitude and altitude, so make sure that you are clear which is which before the big day.

Oceanic circulation

Oceanic circulation is closely associated with general atmospheric circulation and large-scale movements of warm water are part of the process of the horizontal transfer of heat from the tropics polewards. Cold currents also occur, and these carry colder water from the high latitudes towards the equator. Generally, ocean currents are set in motion by the prevailing surface winds. In temperate regions, warm ocean currents act as a moderating influence on the climate, e.g. the North Atlantic Drift causes winter temperatures in coastal areas of western Europe to be higher than those inland. Conversely, a cold ocean current, such as the Labrador Current which flows along the east coast of Canada, gives rise to more extreme winter temperatures.

Knowledge check 9

Identify the main factors influencing global temperatures.

Knowledge check 10

What is the global relationship between aridity and continentality?

Altitude

As a general rule temperatures decrease with height above sea level. In the atmosphere this change is called the **environmental lapse rate**. The average value for this change is 6.5°C for every 1,000 m altitude.

The climate of the British Isles

Climate is defined as the average long-term weather conditions for an area; weather refers to the current day-to-day conditions of the atmosphere at the local scale. The British Isles experience the characteristics typical of the **cool temperate western maritime** climate.

The predominant influences on this climate are:

- its latitude — the British Isles occupy the temperate latitudes between 50° and 60° north of the equator, so day length varies significantly between the seasons and the sun's rays are less intense than in the tropics
- its western aspect — the British Isles is located on the western seaboard of Europe, close to the path of the North Atlantic Drift, a warm ocean current that has a moderating influence on its climate
- its maritime position — adjacent to the Atlantic Ocean, which supplies huge quantities of moisture into the atmosphere

Examiner tip

Revise the typical weather conditions associated with the 5 main air masses affecting the British Isles using a sketch map, identifying the source of each air mass and the nature of the weather relating to each of these.

Knowledge check 11

Distinguish between weather and climate.

Table 4 Characteristics of the cool temperate western maritime climate

Characteristic	Cool temperate western maritime climate
Temperature	Winters mild (average 2–7°C), summers warm (average 15–19°C). Low annual range (10–15°C), increasing from west to east
Precipitation	Occurs all year round. Frontal, convectional and orographic, between 500 and 2,500 mm annually, decreasing from west to east
Winds	Prevailing winds lie in the belt of southwesterlies
Pressure systems	Frequent low-pressure weather systems particularly during spring and autumn, however high-pressure anticyclonic systems do occur in winter and summer
Air masses	Mostly tropical maritime and polar maritime; these meet at a polar front, creating low-pressure weather systems/depressions. Also intervals of tropical and polar continental and arctic maritime air

Knowledge check 12

What types of air mass are associated with the following weather in the British Isles?

- heavy snow in winter
- heatwave conditions in summer

Examiner tip

You need to know about the origin and nature of depressions. You should also be able to recognise a depression on a synoptic chart and must learn the sequence of weather associated with the passing of a low-pressure weather system.

Low-pressure weather systems: depressions

- A depression is an area of relatively low atmospheric pressure and is recognised on a weather map by a system of closed isobars with pressure decreasing towards the centre.
- Isobars are usually close together, producing a steep pressure gradient. This creates strong winds which flow anticlockwise and inwards in the northern hemisphere around the centre of the depression.
- Most depressions affecting the British Isles form in the Atlantic Ocean. Here cold, dense polar air meets relatively warm, light air originating from the tropical latitudes.
- The zone of contact is known as the polar front. The warm air initially rises above the cooler dense air and, as it does so, condensation occurs, resulting in a belt of clouds and precipitation along the front.
- Depressions are associated with gales and precipitation and occur throughout the year.
- Areas in the southern British Isles will often experience a change of wind direction from south to southwest to west to northwest as the depression passes through; the wind is said to **veer**.
- Areas in the northern British Isles will often experience a change of wind direction from southeast to east to northeast to north; the wind is said to **back**.

Table 5 Weather changes associated with the passing of a depression

Weather element	Cold front			Warm front		
	In the rear	At passage	Ahead	In the rear	At passage	Ahead
Pressure	Continuous steady rise	Sudden rise	Steady or slight fall	Steady or slight fall	Fall stops	Continuous fall
Wind	Veering to northwest, decreasing speed	Sudden veer, southwest to west. Increase in speed, with squalls	Southwest, but increasing in speed	Steady southwest, constant	Sudden veer from south to southwest	Slight backing ahead of front. Increase in speed
Temperature	Little change	Significant drop	Slight fall, especially if raining	Little change	Marked rise	Steady, little change

Weather element	Cold front			Warm front		
	In the rear	At passage	Ahead	In the rear	At passage	Ahead
Humidity	Variable in showers, but usually low	Decreases sharply	Steady	Little change	Rapid rise, often to near saturation	Gradual increase
Visibility	Very good	Poor in rain, but quickly improves	Often poor	Little change	Poor, often fog/mist	Good at first but rapidly deteriorating
Clouds	Shower clouds, clear skies and cumulus clouds	Heavy cumulonimbus	Low stratus and strato-cumulus	Overcast, stratus and stratocumulus	Low nimbostratus	Becoming increasingly overcast, cirrus to altostratus to nimbostratus
Precipitation	Bright intervals and scattered showers	Heavy rain, hail and thunder-storms	Light rain, drizzle	Light rain, drizzle	Rain stops or reverts to drizzle	Light rain, becoming more continuous and heavy

High-pressure weather systems: anticyclones

- Anticyclones are generally larger than depressions (up to 3,000 km across) and are dominated by subsiding air, which produces warming and a decrease in relative humidity.
- They are recognised on a weather map by a system of closed isobars with pressure increasing towards the centre. Isobars are usually far apart, winds flow slowly clockwise around the high in the British Isles from the centre outwards.
- Because they are usually slow-moving and stable weather systems, once established they may persist for several weeks.
- In summer they bring welcomed hot sunny conditions to the British Isles, with light winds and little cloud or rain, although clear skies at night can result in temperature inversions that produce dew, mist and coastal fogs.
- In winter these stable conditions favour the development of fog and frost, and pollution may become trapped in the lower layers of the atmosphere by the inversion.

Storm events: their occurrence, their impact and responses to them

Severe gales are associated with deep depressions in mid-latitudes experiencing the cool temperate western maritime climate. The strongest winds occur as a warm or cold front passes over. Generally speaking, the lower the air pressure in the centre of the depression, the steeper the pressure gradient and the higher the wind speed.

Knowledge check 13

How do the impacts on people of an anticyclone vary between winter and summer?

Examiner tip

You must be able to recognise a high-pressure weather system or anticyclone on a synoptic chart and must learn the typical characteristics of the weather in summer and winter.

Examiner tip

You need to know in detail one case study of a significant storm event from the last 30 years, with specific reference to the weather conditions experienced and the impacts and consequences of this event.

Knowledge check 14

How do the responses to a storm event in the British Isles compare with those to a summer drought?

Examiner tip

You need to know the characteristics of one climate type. If you have studied Ecosystems as another option, it would make good sense to learn the climate type corresponding to the required case study of a biome in that option.

Knowledge check 15

Distinguish between the annual and diurnal temperature range.

The climate of one tropical region (tropical wet/dry savanna or monsoon or equatorial)

Basic climatic characteristics; temperature, precipitation and winds

Equatorial climate

- **Temperature** — average monthly temperatures are between 26°C and 28°C. In locations further from the equator a dual peak of temperature may be evident, as the sun is directly overhead twice a year. The diurnal temperature range is 17°C, which is greater than the annual temperature range. Temperatures are lower than in other tropical climates due to extensive cloud cover.
- **Precipitation** — precipitation is convectional, and often comes during the late afternoon. Annual precipitation totals are often over 2,000 mm, but this may vary as a short dry season can be experienced a few degrees away from the equator.
- **Winds** — although the southeasterly and northeasterly trade winds converge at the inter-tropical convergence zone (ITCZ), surface winds are generally light and variable, allowing land and sea breezes to develop in coastal areas.
- **Pressure systems** — the equatorial climate lies within the 'Doldrums' low-pressure belt, where convergence of the trade winds at the ITCZ results in vertical uplift.
- **Air masses** — equatorial maritime or equatorial continental throughout the year.

Tropical wet/dry savanna climate

- **Temperature** — for most of the year temperatures are high. During the hot season average monthly temperature may reach 32°C, in the so-called cool season temperatures are still above 20°C.
- **Precipitation** — seasonal, during the wet season heavy convectional rains bring over 500 mm of precipitation, the dry season increases in length with distance from the equator.
- **Winds** — trade winds are dominant, blowing from the southeast in the southern hemisphere and from the northeast in the northern hemisphere. These are stronger during the dry season.
- **Pressure systems** — low pressure dominates during the wet season and high pressure during the dry season, due to the movement of the ITCZ. In the northern hemisphere the wet season begins in April.
- **Air masses** — tropical continental air dominates during the dry season and alternates with equatorial air during the wet season.

Tropical monsoon climate

- **Temperature** — temperatures range from 32°C in the hot season to around 15°C in the cool season.
- **Precipitation** — seasonally heavy, averaging 2,500 mm but varying. For example up to 10,000 mm on the southwest-facing hillslopes of Assam and much lower inland, e.g. Delhi 620 mm.

- **Winds** — seasonal reversal of winds is the key characteristic of this climate. During the wet season the winds blow from the sea to the land (from the southwest in the Indian subcontinent) and vice versa (from the northeast during the dry season).
- **Pressure systems** — low pressure during the wet season. In the Indian subcontinent, this occurs from June to October. High pressure then dominates for the rest of the year when the ITCZ moves away.
- **Air masses** — tropical continental air dominates in the dry season, equatorial maritime air during the wet season.

The role of subtropical anticyclones and the ITCZ

The global circulation model helps to explain the seasonal pattern of climate experienced within the tropical regions. A satellite photograph of the world reveals the inter-tropical convergence zone (ITCZ) as a band of clouds encircling the centre of the Earth. It is a zone of converging, unstable air, which rises where the trade winds from the northern and southern hemispheres meet. The ITCZ is associated with an area of low pressure and precipitation that moves north and south of the equator throughout the year, following the apparent movement of the overhead sun. As the ITCZ moves it carries with it the low pressure and associated precipitation, allowing anticyclonic conditions to develop in its wake. The rising air at the ITCZ forms the first part of the Hadley cell, which helps to transfer heat globally from the equator polewards. The descending limbs of the two Hadley cells occurring at around 30° north and south of the equator create the subtropical high-pressure belts, where skies are clear and weather conditions are stable and dry.

Knowledge check 16

What is the ITCZ? Why does it have such an important influence on tropical climates?

Tropical revolving storms

Tropical revolving storms are seasonal events and occur during the season dominated by low pressure. They are known as hurricanes in the Caribbean and southeastern states of the USA, cyclones in southeast Asia, typhoons in China and Japan, and willy-willies in Australia.

In summary:

- Tropical revolving storms develop from troughs of low pressure in the easterly trade wind belts and are up to 500 km across. At the centre of the storm is the eye, a central vortex which is an area of calm surrounded by high winds.
- They develop between latitudes 5° and 20° from the equator, where the Coriolis force is able to exert a spinning effect on the rising air.
- Sea temperatures must be over 27°C so that large amounts of water vapour can evaporate, condense and release latent heat.
- The severity of a tropical revolving storm can be categorised using the Saffir-Simpson scale: at scale 5 air pressure is lower than 920 mb and wind speeds exceed 250 km h^{-1}.
- Severe storms of this nature have strong winds, intense precipitation and can cause coastal storm surges, flooding and landslides.
- Once a storm reaches land or cooler latitudes it dissipates.

Examiner tip

You need to know two recent contrasting case studies, including the impacts of, and responses to, each storm. Ensure that you select one from the developed and one from the developing world. It is advisable to categorise the impacts as economic, social, political and environmental; use a table to do so.

Examiner tip

As with depressions, you should revise the sequence of weather experienced as a tropical revolving storm passes over an area. This can be usefully shown on a diagram and might be easier to learn in this way.

Examiner tip
You need to be aware of the changes in temperature, precipitation including thunderstorms, fogs, winds and air quality in urban areas, and to know and understand how and why these variations occur.

Examiner tip
You need to know the human activities responsible for the formation of particulate pollution and photochemical smog, and know how the problem of poor air quality is being tackled by pollution reduction policies in at least two different cities.

Knowledge check 17
Define the terms 'urban heat island' and 'Venturi effect'.

Knowledge check 18
Identify a positive relationship between distance from the city centre and one aspect of the urban climate.

Climate on a local scale: urban climates

Small-scale variations to climate can be seen in urban areas. The larger the city the more marked the effects (Table 6).

Table 6 Effects of urban areas on local climate

	Cause	Effects
Temperature	Solar radiation absorbed and released from buildings; central heating and air-conditioning energy released from buildings	Highest temperatures within the city centre; best observed under anticyclonic weather conditions, just before dawn; less frost in winter
Precipitation	More convection due to heating; more particulate matter to act as condensation nuclei	Higher in large urban areas and downwind of cities; thunderstorms more frequent due to convection; less snow but more fog
Winds	Sheltering effect at ground level from buildings, rougher urban landscapes can cause turbulence due to buildings of different heights; modern streets create narrow passages for wind to pass through	Lower velocities overall, but more turbulence; increase in funnelling of wind, creating local channels of very strong air movement
Air quality	More particulate matter and pollution from vehicle exhausts, factories, power stations and domestic fuel combustion; sunlight reacts with emissions to create ozone	Poor air quality, smog in many cities in newly industrialising countries, photochemical smog in cities under the influence of high-pressure weather systems

Global climate change

Evidence for climate change over the last 20,000 years

Geological evidence shows that change has always been a feature of the Earth's climate. The last major ice age occurred during the Pleistocene period, which ended around 10,000 years ago. During the Pleistocene period, glacial periods occurred cyclically, interspersed with warmer spells, so global warming is nothing new (Table 7).

Table 7 Climatic periods since the Pleistocene ice age

Climatic period	Time before present (years)	Climatic conditions
Sub-Atlantic to present day	2,500	Temperatures fluctuate; cooler than the present day A cool coastal climate with cooler summers and increased rainfall A marked cool period between AD 1300 and AD 1800 — the little ice age A period of warming in the last 200 years
Sub-Boreal	5,000	Temperatures falling but rainfall relatively low at the beginning of this period, increasing later Period known as the neoglacial in Europe, with evidence of ice advance in alpine areas Warm summers and colder winters

Climatic period	Time before present (years)	Climatic conditions
Atlantic	7,500	Temperatures reach the optimum for many trees and shrubs — 'the climatic optimum' A warm 'west coast' type of climate, with higher rainfall
Boreal	9,000	Climate becoming warmer and drier A continental-type climate
Pre-Boreal	10,300	Mainly cold and wet, but becoming warmer and drier Changing from tundra/sub-arctic to more continental

Evidence for past climate change is mostly indirect and has been found in:

- pollen analysis
- dendrochronology (analysis of tree rings)
- ice-core analysis
- sea-floor analysis
- radiocarbon dating
- Coleoptera (remains of beetle species)
- changing sea levels
- glacial deposits
- historical records

Knowledge check 19

What evidence exists to show that climate change is a natural phenomenon?

Examiner tip

You should understand how several of these can provide evidence of climate change.

Recent evidence

According to scientists, in recent years the rate of global warming has exceeded anything seen before. Accurate records of temperature have been kept since the 1800s and indicate that average global temperatures have risen by just under 1°C over the last century, and for the past 30 years temperatures have been warming by some 0.2°C per decade. At present the Earth is as warm as it was during its peak in the last interglacial period, when sea levels were estimated by geologists to have been at least 25 m higher than they are today.

Causes of global warming

- Global warming is a consequence of the greenhouse effect. The greenhouse gases (water vapour, carbon dioxide, CFCs, methane and nitrogen, see Table 8) absorb long-wave radiation from the Earth and prevent it from escaping into space.
- The accumulation of greenhouse gases in the atmosphere has resulted in a rise in temperatures throughout the world in recent decades.
- Evidence suggests that this process is accelerating as emissions of gases continue as a result of human activity.

Knowledge check 20

What is the greenhouse effect?

Table 8 Sources of greenhouse gases

Greenhouse gas	Source
Water vapour	Evaporation and transpiration within the hydrological cycle
Carbon dioxide	Respiration, combustion of fossil fuels, deforestation and burning of vegetation
Methane	Pastoral farming, paddy rice cultivation, decaying vegetation and landfill emissions
Ozone	Chemical reaction between vehicle emissions and sunlight
Nitrous oxides	Agricultural fertilisers, vehicle exhausts
Chlorofluorocarbons (CFCs)	Aerosols, refrigerators (although greenhouse gas CFCs are associated with ozone depletion and have only a very small influence on global warming)

The effects of global warming

Examiner tip

You need to know about the predicted effects of global warming on a global scale, on the British Isles and on the tropical region studied (equatorial or savanna or monsoon climatic region).

The impacts of global warming are open to debate, and depend largely on the computer model used. Globally, as the climate warms, the following impacts are likely:

- Ice caps and glaciers are expected to melt, and sea levels to rise. This may lead to the permanent loss of some low-lying coastal areas and islands, e.g. the Maldives.
- Global warming is predicted to have a number of effects on the oceans. The present global circulation of ocean currents may be disrupted; some scientists also predict the acidification of the oceans and others widespread bleaching and loss of coral.
- Patterns of agriculture are expected to change as some areas become wetter and some drier. Cultivation may cease in large areas of Africa if the Sahara desert spreads in extent. This could lead to global food security issues.
- Extreme weather events are likely to be more common occurrences and may be experienced on a much greater scale too.
- Tropical diseases, such as malaria, may be experienced over a much wider area.
- Globally, biodiversity is likely to shrink as large numbers of plant and animal species become extinct.

Some of the possible impacts on the British Isles and Bangladesh (within the tropical monsoon climate belt) are set out in Table 9.

Table 9 Possible impacts of global warming

British Isles	Bangladesh
Warmer and wetter; seasonal rainfall increases in intensity, southern Britain more likely to experience a Mediterranean climate although the north and west could be much wetter. Storms may increase in frequency and intensity. Some predict colder winters	Increase in total precipitation and length of the wet season with more frequent tropical cyclones. Overall temperatures could increase throughout the year but the wet monsoon could become more unreliable
Coastal flooding increases as sea levels rise, rivers are also expected to flood more frequently and flooding events are predicted to worsen in magnitude. More investment in coastal and river basin management, otherwise population will be displaced	Increased flooding as sea levels rise, causing widespread displacement of population towards India. This will create international tension
Changes in agricultural practices and crops, e.g. maize, vines and other Mediterranean produce. More irrigation may be necessary in some areas	Longer wet season may increase the length of the growing season and allow multi-cropping of paddy rice
Water supply may become a problem in some areas during the summer, leading to restrictions	The monsoon may become less reliable and drought may occur, causing a water supply crisis
Increases in pests and diseases e.g. malarial mosquitoes may spread to the British Isles	Many species may become extinct as their habitat is lost, e.g. Bengal tiger

Responses to global warming

International

- The **Earth Summit** in 1992 took place in Rio de Janeiro, Brazil. This was a meeting of most of the world's countries and an agreement called Agenda 21 was passed. One of its aims was to cut environmental pollution to conserve resources and protect natural habitats and wildlife.
- The **Kyoto Protocol** was an international agreement signed by more than 100 countries in Kyoto, Japan in 1997, the aim of which was to halt climate change. Countries made pledges to cut their carbon emissions by agreed amounts by 2010. By 2006 it had been ratified by 162 countries, but the USA was criticised for refusing to adhere to it. This treaty is due to expire in 2012.
- The **Copenhagen Conference** of 2009 was set up by the United Nations to put in place a follow-on agreement to Kyoto. Following the election of its new president, Barack Obama, in 2009, the USA appeared to be more committed to cutting emissions. However, it proved impossible to reach a global agreement at this meeting. The **'Copenhagen Accord'** was not ratified unanimously, so there is currently no binding global pact in place to cut emissions. In 2011 more than 130 countries had made varying commitments to cut emissions by 2020. These amounts differ tremendously — for example, Norway pledged cuts of 40%, comparing favourably with the USA's commitment to reduce emissions by just 4%.

Examiner tip

You need to be aware of the strategies currently employed to reduce the impact of global warming at the international, national and local scales, and be able to assess the effectiveness of such strategies.

Examiner tip

You need to know about the recent and future international political responses to global warming, including the Rio Earth Summit, the Kyoto Protocol and the Copenhagen Accord.

British Isles

The 2006 Climate Change Programme set out strategies to address the issue of climate change and the 2008 Climate Change Bill attempted to put these into practice. The UK government set a target to reduce the output of carbon dioxide by 20% before 2010, cutting emissions by at least 34% by 2020 and 80% by 2050 (below the 1990 baseline). It has also committed to increasing its renewable energy sources, in particular wind energy, and has decided to re-invest in nuclear power to help cut emissions. It also aims to improve the provision of recycling, to expand the production of biofuels and to encourage the overseas purchase of 'carbon credits'. Building regulations have been tightened to ensure that new homes are carbon efficient.

Local scale

Local governments are encouraging a 'think globally, act locally' approach and many are investing in, and encouraging the use of, public transport. The Greater London Authority, under its former leader Ken Livingstone and latterly Boris Johnson, has attempted to tackle this issue. London has a congestion charge and a low emission zone in place to try to restrict the volume of traffic travelling through the city centre. Recently, a successful bike hiring scheme has been been set up in London, sponsored by Barclays Bank. Cities such as Sheffield and Nottingham have re-introduced electric trams, which are more environmentally friendly than diesel-burning buses. Most local councils have recycling schemes and households are encouraged to sort their refuse to cut down the amount of landfill. Education programmes are also being implemented by local education authorities to encourage sustainable energy use and to cut down on waste.

Summary

Weather and climate and associated hazards

- The world's climates are the outcome of a wide range of planetary controls. These involve the structure of the atmosphere, the heat budget, the general atmospheric circulation and the systems of oceanic circulation.
- The climate of the British Isles is the outcome of the wide variety of air masses that affect it, which in turn are brought to these Isles by depressions and anticyclonic conditions. The British Isles also experience significant storm events on a regular basis.
- Tropical regions also experience their own climates, and these vary within the tropics. They include equatorial climates, monsoon climates and savanna climates. The roles of the subtropical anticyclones and inter-tropical convergence zone are crucial in creating each of these climates.
- Tropical regions also experience significant storm events — known as tropical revolving storms (cyclones and hurricanes). These can affect both developed and developing countries.
- Urban areas can modify the prevailing climates that affect them in terms of temperature, precipitation, winds and air quality.
- Evidence suggests that the world is experiencing a general warming of temperatures, commonly referred to global warming or climate change. There is a range of views on the causes of this warming, and its possible effects. Various responses at a range of scales have been put in place to counteract it.

Ecosystems: change and challenge

Nature of ecosystems

An **ecosystem** is defined as a living system of plants and animals which interacts with the physical environment. Ecosystems can be considered at any scale, from a small area such as a pond to an area as large as the Earth itself.

Knowledge check 21

Distinguish between an ecosystem and a biome.

A **biome** is defined as an ecosystem on a global scale, with a climax community of plants and animals which has reached equilibrium with its environment. Within a biome the vegetation and soil are predominantly influenced by the climate.

Structure of ecosystems

An ecosystem consists of biotic (living) and abiotic (non-living) components.

Knowledge check 22

What is the difference between the biotic and abiotic components of an ecosystem?

Biotic components:

- vegetation
- animals, birds, insects and other living creatures
- micro-organisms, such as plankton
- decomposers

Abiotic components:

- climate — rainfall, temperature, sunlight
- soil characteristics
- underlying parent rock
- relief of the land
- drainage characteristics
- water — fresh or salt

Energy flows, trophic levels, food chains and webs

Energy flows through an ecosystem from one trophic level to another. A trophic pyramid diagram shows the feeding levels that operate within an ecosystem. Initially, energy from the sun allows photosynthesis to take place, and plants or primary producers form the first trophic level. The next trophic layer (primary consumers or herbivores) feeds on the plants but only a small proportion of energy is transferred from one level to the next. Around 90% of energy is lost from each layer through life processes, such as movement, growth, respiration, reproduction and excretion. The secondary consumers or carnivores lie above the herbivores and the tertiary consumers or top predators lie at the top of the pyramid. Detritivores and decomposers operate at each level.

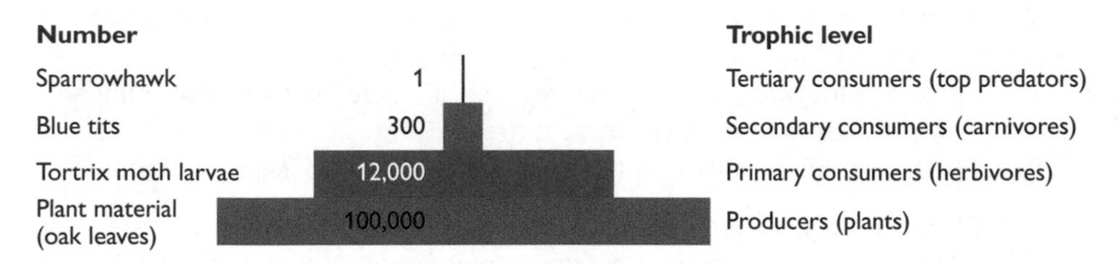

Figure 3 An energy pyramid

The movement of energy up the trophic levels demonstrates the food chain, a simple way of describing the order in which species feed on each other. An example is shown in Figure 3. In reality, relationships are more complex as some species can occupy more than one position; for example a caterpillar is eaten by a variety of bird species that in turn are the prey for a number of different species of predator. A food web diagram can be used to show a larger number of food chains operating within a single ecosystem.

Knowledge check 23

How is energy lost between trophic levels?

Nutrient cycling of minerals within an ecosystem between the abiotic and biotic components can be demonstrated using a Gersmehl diagram (Figure 4):

- nutrients are held in three stores — litter, biomass and soil
- nutrients are transferred between the stores by leaf fallout, decomposition of litter and uptake of nutrients from the soil by plants
- inputs of nutrients include minerals dissolved in precipitation and from weathering of the parent rock
- outputs from the nutrient cycle include losses in runoff and leaching

Knowledge check 24

What does a Gersmehl diagram show?

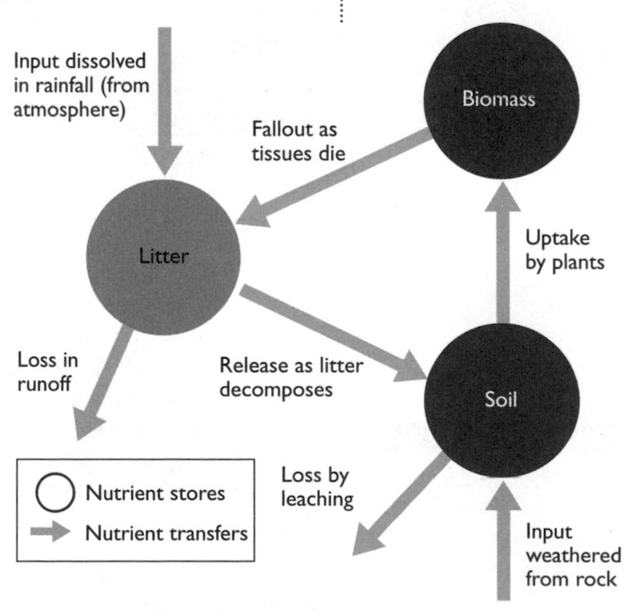

Figure 4 A model of the mineral nutrient cycle

Ecosystems in the British Isles over time

Succession and climatic climax

Examiner tip

You need to know and to understand thoroughly the sequence of seres relating to one type of plant succession found in the British Isles. A labelled diagram could be used in the exam to illustrate all the stages.

- Time is an important factor in the development of ecosystems. A sere is a community of plants that forms a stage in the development of vegetation over time.
- The complete sequence of seres from the initial colonisation by pioneer communities to climatic climax is known as a primary succession or prisere.
- The pioneer community is the first stage in a succession. Here hardy plants, mosses and lichens adapted to survive in tough conditions start to colonise.
- The pioneer community is succeeded in time by a more complex variety of species as the growing conditions improve.
- The climatic climax sere eventually develops. Here the plants and animals become a stable community, in balance with their environment.
- In the British Isles four basic types of prisere exist; lithosere (bare rock), psammosere (sand), halosere (salt water) and hydrosere (fresh water).
- A typical lithosere might consist of the following sequence of seres or stages:

> Algae/bacteria–lichens–herbs/grasses/small flowering species–ferns/brackens/brambles–large shrubs/small trees–large trees, e.g. oak and ash

Knowledge check 25

Define the terms prisere and climatic climax.

The temperate deciduous woodland biome

Examiner tip

In the exam, if presented with a photograph of a deciduous woodland you must use this to explain what is actually there, not aspects, such as brown earth soil, that you can't actually see.

Temperate deciduous woodland is an example of a high-energy biome, with high productivity in relation to other global vegetation zones. It is found in the mid-latitudes, where there is adequate moisture; it is not found in the interiors of continents where patterns of climate are more extreme. In the British Isles oak was originally the dominant species in the lowlands, although ash was common in some areas.

Climate

- Average winter temperatures 2–7°C, summers 13–17°C.
- Total precipitation 500–2,000 mm y^{-1}, throughout the year with a winter maximum.
- On-shore westerly winds predominate, moderating temperatures and bringing moist air.
- Low pressure weather systems dominate.

Vegetation

- Broadleaved deciduous trees, such as oak, ash, beech and birch, shed their leaves in autumn, before the onset of colder winter temperatures.
- As soil temperatures fall the tree roots can only absorb small amounts of water so growth is retarded.
- Heat loss reduces transpiration and the demand for water during the cooler months.

Knowledge check 26

Why is deciduous woodland considered a high-energy biome?

Soil

- Brown earth, a fertile zonal soil approximately 1.5 m deep, dominates.
- Leaf litter accumulates during autumn and is quickly decomposed the following year by a wide range of soil organisms.

- The soil is well-mixed by earthworms and other soil organisms so layers are indistinct.
- Mild leaching occurs, particularly during autumn and winter when precipitation exceeds evaporation.

The effects of human activity on plant succession

People interfere with plant succession. Interfering or **arresting factors** are those that stop a plant community from reaching its climatic climax. If a break in succession is maintained it is known as a **plagioclimax**. Sheep grazing on moorland and preventing it developing into woodland would be an example. A **secondary succession** is one that re-starts on land that has previously been vegetated. Secondary succession can occur following a natural event, such as a volcanic eruption, or can follow a period of human interference, such as the re-colonisation by vegetation of a former quarry site.

Originally much of the British Isles was covered in deciduous woodland, with the climatic climax either oak or ash, depending on the underlying parent rock. By the Middle Ages most of this had been cleared for agriculture and settlement. In recent decades pollution from the industrial heartland of Europe has resulted in acid rain, which has had a significant impact on vegetation and aquatic ecosystems in northwest Europe.

Knowledge check 27

What is the difference between a secondary succession and a plagioclimax?

Examiner tip

You need to know about the characteristics of one plagioclimax, for example heather moorland typical of a British upland area such as the Pennines. You should be able to demonstrate an understanding of the changes that have taken place.

The biome of one tropical region

You need to study one tropical biome, either tropical savanna grassland or tropical monsoon forest or equatorial rainforest. In particular you should cover:

- *the main characteristics of the climate, vegetation, soil and the biodiversity of the wildlife*
- *the way in which the vegetation and animals have adapted to the climate and soil*
- *human activity and its impact on the biome, including the causes and consequences of deforestation and/or desertification*

The equatorial rainforest biome

Climate

- Little seasonal variation in climate, low pressure dominates through the year.
- Mean monthly temperatures 25–28°C (diurnal range greater than annual range in temperatures).
- Humidity remains high throughout the year.
- Annual precipitation often over 2,000 mm, evenly distributed, although there might be a short dry season away from the equator.
- Convectional rain occurs most afternoons.
- Both day and night are roughly 12 hours long throughout the year.

Vegetation

- Evergreen appearance due to year-long growing season.
- Forest has a five-layer structure, tallest trees called emergents are up to 45 m tall.
- Upper and lower canopy layers provide continuous cover.

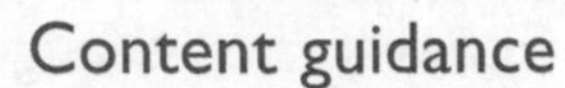

- Tallest trees have developed buttress roots to support their great height.
- Leaves have drip-tips to help shed rain.
- Trunks are branchless under the canopy, where it is too dark for photosynthesis.
- Plants such as lianas grow on trees as there is insufficient light on the forest floor.
- Tree roots are generally near or on the surface as soil is nutrient-deficient deep down.

Soils

- Latosols or ferralitic soils dominate — these are red-coloured, nutrient-poor soils.
- Soils are up to 40 m deep — resulting from rapid chemical weathering (ferralitisation) due to hot wet climate.
- Leaching/eluviation occurs due to moisture surplus; evapo-transpiration exceeds precipitation.
- Red colour from the accumulation of iron and aluminium oxides.
- Nutrient-poor, thin humus layer despite rapid decomposition of litter due to equally rapid uptake of nutrients from soil by vegetation.

Examiner tip

Ensure that you can discuss the problems relating to sustainability within the biome studied and whether maintaining this is compatible with economic development.

Biodiversity

- Most diverse and productive biome on Earth.
- Up to 300 species of tree per square kilometre, including mahogany, teak, rosewood and brazil nut.
- Amazon rainforest has nearly 600 species of mammal, nearly 2,000 species of birds, more than 1,500 species of amphibians and fish.
- Scientists believe that some species are as yet undiscovered.
- Threatened species include gorilla, chimpanzee and orang-utan, jaguar, parrot and toucan.

The tropical savanna grasslands

Climate

- Marked wet and dry seasons, length of dry season increases with distance from the equator.
- Annual rainfall ranges from 500 mm to 1,000 mm y^{-1} depending on distance from the equator (lower further away).
- Hot throughout the year, average monthly temperatures range from 18° to 28°C, double peak in temperatures may occur as sun is directly overhead twice per year.
- Dry season dominated by subtropical high pressure and trade winds.
- Wet season dominated by ITCZ and low pressure.
- Spontaneous fires may occur, started by lightning.

Vegetation

- Trees dominate over grasses where the wet season is longer.
- Grasses dominate in locations where the dry season is longer, further away from the equator.
- Grasses may be higher than 2 m tall. Their long roots reach down to underground moisture.
- Both grasses and trees are deciduous, losing their leaves in the dry season.

- Trees have adapted to survive drought, e.g. baobabs store water in their swollen trunks to survive the dry season.
- Trees develop xerophilous features: deep, branched roots seek moisture underground.
- Evergreen trees also occur, with hard leathery leaves to reduce transpiration loss.
- Acacia trees have developed flattened crowns, to cope with strong trade winds.
- Vegetation has adapted to cope with fire.

Soil

- Deep red lateritic soils, with a hard cemented crust of iron/aluminium oxides below the surface.
- The soil is influenced by the seasonal climate, in particular the rainfall.
- During the dry season soil solution rises by capillary action, carrying and then depositing minerals and forming the crust.
- During the wet season silica is removed from the upper layers of the soil and leached downwards.

Biodiversity

- A greater diversity of vegetation and wildlife exists towards the equator.
- Tree species include acacia, baobab, umbrella thorn; grasses include elephant and other tall species.
- Within the Tanzanian savanna biodiversity is lower than in the rainforest, with some 375 species of mammal, around 1,000 species of birds and some 800+ species of reptiles, amphibians and fish.
- Great animal migrations occur within this biome as wildlife are forced to move to seek water when the dry season arrives.
- Almost 40% of the land in Tanzania has been designated as National Park to help protect biodiversity.
- Endangered species include the African elephant, the black rhino, zebra and cheetah.

Examiner tip

Learn a Gersmehl diagram to show nutrient cycling within the biome that you have studied.

The tropical monsoon forest biome (e.g. Indian subcontinent)

Climate

- High temperatures throughout the year (19–30°C), maximum when sun is directly overhead.
- Seasonal reversal of winds due to movement of ITCZ.
- During winter northeast winds blow from the interior of Asia, high pressure dominates and dry, sunny conditions prevail.
- In the summer months winds blow in from the SW, over the Indian Ocean bringing low pressure and heavy rainfall.
- Total annual precipitation varies (500–2,500 mm+) depending on location in relation to the coast. Inland much lower.

Vegetation

- The indigenous vegetation has been largely cleared for agriculture in the lowlands.
- Mangrove forest dominates in the coastal lowlands, where vegetation has adapted to cope with saline conditions.

- The trees are lower in height than in the rainforest, providing a less continuous cover.
- Most trees are deciduous, losing their leaves in the dry season.
- Different species all fruit and flower at the same time in line with the seasons.
- The open canopy allows a dense undergrowth of bamboo to develop.

Soils

- Lateritic, characteristics shared with soils in both the savanna grasslands and rainforests.
- Dry season is often shorter than that experienced in the African savanna so soils may reach similar depths to those in the rainforest due to rapid chemical weathering in warm, moist conditions.
- Deltaic soils may be more fertile due to nutrients deposited following frequent river floods.

Biodiversity

- The most continuous untouched monsoon forest remains in Myanmar, Thailand, Laos, Vietnam and Cambodia. More than 90% has been deforested in India.
- Common species of trees and plants include teak, sal, bamboo and orchid.
- Fewer species of mammals, birds, reptiles, fish and amphibians than in tropical rainforest and savanna grassland.
- Species include Bengal tiger, Indian elephant, leopard, rhinoceros, python and cobra.
- Many species are endangered through hunting or loss of habitat.

Ecosystem issues on a local scale: impact of human activity

Urbanisation

The process of urbanisation has occurred over the last 200 years, first in the more economically developed regions such as western Europe and North America and, since the mid-twentieth century, in the newly industrialising and less developed regions of the world, spreading from Asia through to South America and Africa.

Urban land uses have a wide variety of potential ecological habitats for both plants and animals. There are opportunities for secondary plant succession, particularly where land becomes derelict, and in other places a more permanent plagioclimax might exist. Possible environments include:

- industrial sites, some of which may have become derelict over time and some may still be in use
- transport routes, both used and disused, including roadside verges, railway embankments, and canal and river banks
- residential gardens and allotments
- parks and other open green spaces
- landfill sites
- water bodies, such as ornamental lakes and reservoirs
- pockets of urban woodland

Knowledge check 28

Why did de-industrialisation result in secondary succession?

Urban niches

An urban niche is a small-scale area or micro-habitat within a town or city, providing a specialised environment within which a certain species of plant can develop and thrive. An example is mosses taking root and growing undisturbed on top of high walls and ledges.

Cities are dynamic environments, constantly evolving and experiencing change. For example, new build is an ongoing process in the rural–urban fringe, with re-urbanisation, dereliction and decay more common in central areas. In many British cities, de-industrialisation has occurred, resulting in significant areas of wasteland where traditional industries once thrived. Old factory sites, docklands, railway sidings and storage depots have been abandoned in many areas, for example London's Docklands, Sheffield's steelworks and Stoke-on-Trent's potteries.

A secondary succession will develop on any wasteland surface:

- concrete/brick/stone will provide the initial surface for a lithosere (bare rock succession)
- gravel pits filled with water and disused canals will allow the development of a hydrosere

On an old industrial site surfaces will not be uniform: some surfaces will be tarmac, some concrete, wetland or rubble. Some sites will be acid, some alkaline, some might be so contaminated that nothing can grow. Such differences provide small-scale urban niches and are called substratum variations where several parallel successions take place.

Routeways, in particular verges of railways and dual-carriageway roads where there are few pedestrians, provide distinctive habitats as exotic species can be brought in by traffic. They also provide wildlife corridors for species such as foxes, comparable with hedges in rural areas. Canals act like ponds, providing a habitat for a variety of aquatic plants and insects.

Exotic species such as buddleia have been introduced into many gardens and have spread into open spaces. Japanese knotweed is a problematic species as it is invasive, resistant to pesticides and difficult to get rid of once it has started to colonise.

Changes in the rural–urban fringe

The rural–urban fringe is the open countryside that is adjacent to the city and is in demand for a variety of land uses including:

- potential space for the development of new housing, business and industrial parks
- transport including roads and airports
- recreation: parks, sports stadia, golf courses
- waste disposal including landfill sites and sewage works
- agricultural land

Land on the urban fringe is often degraded; owners speculate that lucrative planning permission is more likely to be granted if the land is run-down and unkempt. Such land is often affected by problems such as fly-tipping, vandalism and illegal encampment. Untended land will become overgrown by weeds and brambles as secondary succession occurs.

Examiner tip

If you are given a photograph showing vegetation or wasteland in an urban area in the exam ensure that you use this and only describe and explain what you can see, not what you think should be there.

Knowledge check 29

Outline how distinctive plant ecologies develop along roads and railways.

In the British Isles, many cities are surrounded by green belt land, which is protected by law from development. However, the government is planning for 500,000 new homes in the next 25 years so it is likely that such land will be under renewed pressure for development.

Examiner tip

You need to study one example of an ecological conservation area at a local scale. The study does not have to be based in an urban area — but it must be small scale. You must be able to assess the extent to which this area has been successful in achieving its conservation goals.

Ecological conservation areas

There has been increasing pressure to protect the environment in recent decades and there are examples of small-scale conservation projects in most urban areas. Conservation helps to:

- encourage wildlife back into cities
- make cheap use of an otherwise derelict site
- reduce maintenance costs in an area
- maintain a diverse species base and
- re-introduce locally extinct species

Ecosystem issues on a global scale

Human activity, biodiversity and sustainability

Human activity is important because it can modify the natural environment, with both detrimental and positive impacts.

Biodiversity is the variety of different forms of life within an ecosystem. A complex and species-rich ecosystem is considered healthy and desirable. The Earth, a global ecosystem, has remarkable biodiversity with many millions of species that are the result of billions of years of evolution. There has been a remarkable loss in global diversity over the last 200 years due to human activity and it has been predicted that climate change will result in the extinction of thousands of species over the next 50 years.

Sustainability requires the careful management of the world's resources so that future generations can continue to benefit from and enjoy them as we do today. Sustainable development involves the management of resources in such a way that the ability of the system to replace itself is greater than the level of exploitation.

Knowledge check 30

Define the terms fragile environment and sustainable management.

Environmental issues have become more important politically in recent years at both global and national scales, and sustainable management of fragile environments to protect biodiversity is vital. Many global agreements have been passed to help ensure biodiversity, such as the treaties signed at the Convention on Biological Diversity in Kyoto, 1997, and at the World Summit for Biodiversity in Johannesburg in 2002.

Management of fragile ecosystems

A fragile ecosystem is one that lacks resilience to change. A species-rich ecosystem is more resilient to change than one which is species-poor. According to the United Nations, fragile ecosystems include:

- deserts and semi-arid areas, including the African savanna
- mountainous regions
- polar locations
- freshwater and intertidal wetlands, such as the mangrove forests and swamps in Bangladesh

- tropical rainforests such as those in the Amazon basin
- coral reefs

Human activity in its many forms, from deforestation and cultivation to industrialisation and pollution, has put increasing pressure on ecosystems worldwide. Conservation schemes aim to stop the overexploitation of fragile ecosystems by human activity in order to protect the environment and to maintain biodiversity.

Management schemes

You have to study the management of contrasting fragile environments in terms of conservation versus exploitation. Key elements here are 'management', 'fragile' and 'contrasting'.

Examples include the Everglades of Florida, the Okavango swamps in Botswana, the Serengeti National Park, the Sundarbans in Bangladesh and the Comoros Islands; there are many more. Each area chosen should be subject to pressures in terms of development (exploitation) with, at the same time, people wishing to protect it (conservation). Therefore, a balance needs to be struck between the two processes so that the various attitudes towards these can emerge. Questions asked will focus on the processes of management, conservation and exploitation, and the extent to which these processes can exist alongside each other.

Examiner tip

Two contrasting case studies of recent management schemes (within the last 30 years) in fragile environments should be learned. It would be sensible to use a management scheme from the tropical biome you have studied.

Summary

Ecosystems: change and challenge

- All ecosystems can be described in terms of their structure, energy flows and trophic levels. Linked to these are the concepts of food chains and food webs.
- The ecosystems of the British Isles have developed over time, through the natural process known as succession. The climatic climax vegetation is the temperate deciduous woodland. Equally, human activity has had a significant impact on succession, resulting in a number of plagioclimaxes such as heather moorland.
- The tropical regions of the world also have distinctive ecosystems, ranging from the equatorial rainforest to the monsoon forest to the savanna grasslands. Plants and animals in these biomes have adapted to the climatic conditions. Each of these biomes is under pressure from human activity and therefore faces issues associated with development and sustainability.
- At a local scale, distinctive ecosystems can evolve in locations such as wasteland, parks and gardens and along routeways. Some people seek to conserve localised ecologies through the creation of conservation areas.
- A large number of ecosystems around the world are under threat from human activity and development pressures. Such fragile environments need to be well-managed so as to balance the need for conservation with the demands for exploitation.

World cities

The global pattern of urbanisation

At a global scale, rapid urbanisation has occurred over the last 50 years (Figure 5). Fifty per cent of the world's population lives in towns and cities; over 20% of the population lives in cities of over 1 million. The most urbanised continents are Europe,

North America, South America and Oceania; the least urbanised are Asia and Africa. However, in terms of urban growth, the number of urban dwellers is by far the largest in Asia, with 1.8 billion people, 43% of the population, living in towns and cities. A consequence of the rapid economic development that is taking place in parts of China, India and southeast Asia is that levels of urbanisation will increase very rapidly.

Examiner tip

You should be able to describe the pattern of urban living as shown in Figure 5. Which areas/countries have high levels, which have low levels and where are the anomalies to these patterns? You should also be able to relate this pattern to varying levels of development.

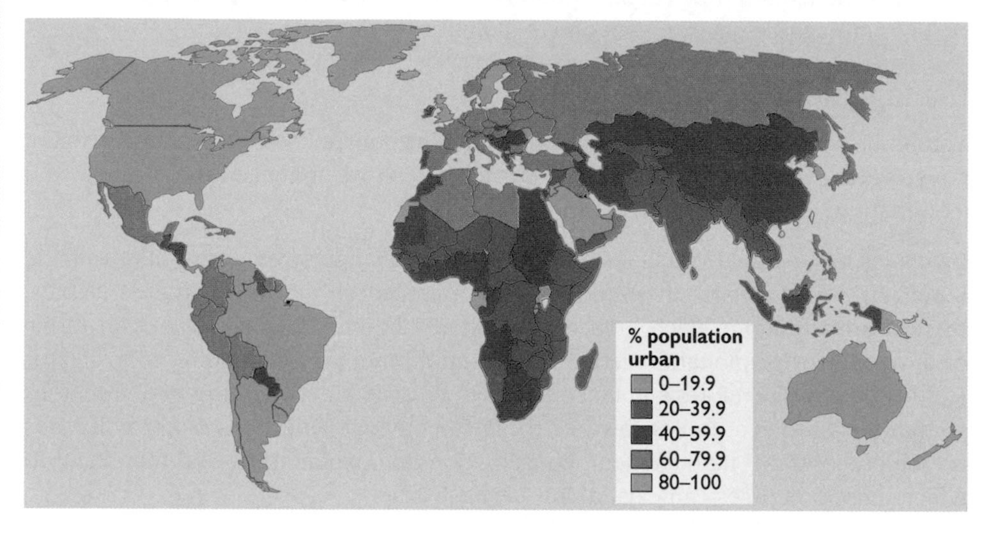

Figure 5 Percentage of the world's population living in urban settlements, 2010

- **Millionaire cities** — Cities with more than 1 million people. India and China have the most millionaire cities in the world.
- **Megacities** — Cities with more than 10 million people, of which, in 2010, there were 28 (18 in the developing world).
- **World cities** — Cities which have great influence on a global scale, because of their financial status and worldwide commercial power.

Examiner tip

You should know the meaning of, and be able to give examples of, these key terms.

Knowledge check 31

Distinguish between megacities and world cities.

For each of the following processes of ***urbanisation***, ***suburbanisation***, ***counter-urbanisation*** *and* ***re-urbanisation****, you should make use of case studies of countries at different levels of development. Also, in each case, make sure that you are aware of how planners and managers in the city have responded to the issues that have arisen.*

Urbanisation

Urbanisation is an increase in the proportion of a country's population that lives in towns and cities. There are two causes of urbanisation:

- Natural population growth — urban areas tend to have relatively low age profiles. Young adults (15–40 years) have traditionally migrated from rural areas. They are in their fertile years and so the rates of natural increase are higher in cities than in the surrounding rural areas.
- Rural–urban migration — the reasons for rural–urban migration are often divided into 'push' and 'pull' factors. **Push factors** cause people to move away from rural areas, whereas **pull factors** attract them to urban areas. In countries with lower levels of economic development push factors tend to be more important than pull factors.

You should be able to discuss both causes of urban growth in the context of at least one city. Detail of the characteristics and the effects of urbanisation (for example, on people, housing types, the environment, quality of life, planning issues etc.) on named areas is also required. Case studies could arise from cities such as São Paulo, Manila, Mumbai and Lagos.

Examiner tip
Make sure you have some pieces of information that could only apply to your chosen city/ies. This enables a strong 'sense of place'.

Suburbanisation

Suburbanisation has resulted in the outward growth of urban development that has engulfed surrounding villages and rural areas. During the mid- to late-twentieth century, this was facilitated by the growth of public transport systems and the increased use of the private car. The presence of railway lines and arterial roads has enabled relatively wealthy commuters to live some distance away from their places of work but in the same urban area. The edge of town, where there is more land available for car parking and expansion, became the favoured location for new offices, factories and shopping outlets. In a number of cases, the 'strict control' of the green belts was ignored (or at best modified) in the light of changing circumstances. More recently, there has been the development of new housing areas on 'previously developed land', leading to the infilling of large areas of private gardens and other open spaces in urban areas, such as school playing fields.

You should be able to discuss the characteristics, causes and effects of suburbanisation on a specific urban area. Many students refer to a UK-based case study here, but there is no requirement to do so. You may also wish to investigate the use of both greenfield (not previously developed) and brownfield (previously developed) sites in the UK.

Examiner tip
Make sure you have some pieces of information that could only apply to your chosen area(s). This enables a strong 'sense of place'.

Counter-urbanisation

Counter-urbanisation is the migration of people from major urban areas to smaller urban settlements and rural areas. There is a clear break between the areas of new growth and the urban area from which the people have moved. As a result, counter-urbanisation does not lead to suburban growth, but to growth in rural areas beyond the main city.

Examiner tip
Be clear about the distinction between suburbanisation and counter-urbanisation.

A number of factors have caused the growth of counter-urbanisation:

- A negative reaction to city life. Many people want to escape from the air pollution, dirt and crime of the urban environment.
- Greater affluence and car ownership allow people to commute to work from such areas. Indeed, many sources of employment have also moved out of cities. Improvements in technology such as the internet have allowed more freedom of location.
- There has been a rising demand for second homes and earlier retirement into rural areas. The former is a direct consequence of rising levels of affluence.
- The need for rural areas to attract income. Agriculture is facing economic difficulties and one way for farmers to raise money is to sell unwanted land and buildings.

Knowledge check 32
Distinguish between urbanisation, suburbanisation and counter-urbanisation.

You should be able to discuss the characteristics, causes and effects of counter-urbanisation on a specific small town or rural area. Many students refer to a UK-based case study here, but there is no requirement to do so. It is important that you are precise about the consequences of the process in the named area.

Examiner tip
Make sure you have some pieces of information that could only apply to your chosen area(s). This enables a strong 'sense of place'.

Re-urbanisation

Re-urbanisation is the movement of people into the city centre or inner city as part of the process of urban regeneration. There are three main processes:

- in-movement by higher-income individuals or groups of individuals into older housing that was in a state of disrepair and the improvement of that housing — **gentrification**
- in-movement by people as part of large-scale investment programmes aimed at urban regeneration in a wider social, economic and physical sense — **property-led regeneration schemes**
- the move towards **sustainable communities** (and partnerships), allowing individuals and communities who live in city centres to have access to a home, a job and a reliable income. Urban social sustainability should provide a reasonable quality of life and opportunities to maximise personal potential through education and health provision, and through participation in local democracies

Knowledge check 33

Distinguish between re-urbanisation and regeneration.

Examiner tip

Make sure you have some pieces of information that could only apply to your chosen areas. This enables a strong 'sense of place'.

You should be able to discuss each of these processes in the context of a range of case studies. It is suggested that you study one area for each process, and that you consider how re-urbanisation and regeneration have taken place in your chosen city in the developing world.

Gentrification

Gentrification is a process of housing improvement. It is associated with a change in neighbourhood composition in which low-income groups are displaced by more affluent people, usually in professional or managerial occupations. Gentrification involves the rehabilitation of old houses and streets on an individual basis, but is openly encouraged by groups such as estate agents, building societies and local authorities. The purchasing power of the residents is greater, which leads to a rise in the general level of prosperity, and there is an increase in the number of bars, restaurants and other higher-status services. The very nature of the refurbishment that takes place in each house leads to the creation of local employment in areas such as design, building work, furnishings and decoration.

Examiner tip

It is always better to describe this process in terms of an actual small-scale area that you have studied.

Property-led regeneration

Urban Development Corporations (**UDCs**) were set up in the 1980s to take responsibility for the physical, economic and social regeneration of selected inner-city areas that had large amounts of derelict and vacant land. They were given planning approval powers over and above those of the local authority, and were encouraged to spend public money on the purchase of land, the building of infrastructure and on marketing to attract private investment. The appointed boards of UDCs, mostly made up of people from the local business community, had the power to acquire, reclaim and service land prior to private-sector involvement and to provide financial incentives to attract private investors.

There were two significant criticisms of UDCs. First, they were too dependent on property speculation and they lost huge sums of money through the compulsory purchase of land which subsequently fell in value. Second, because they had greater powers than local authorities, democratic accountability was removed. Local people often complained that they had no involvement in the developments taking place.

Examiner tip

Be clear of the distinction between property-led schemes and partnership schemes in urban regeneration.

Sustainable communities

Sustainable communities exist in a variety of UK towns and cities and endeavour to:

- have decent homes for sale or rent at a price people can afford (often known as 'affordable housing')
- safeguard green spaces that are well-designed
- be effectively governed with a strong sense of community

Urban decline and regeneration

Urban decline

Inequalities occur in all urban areas. Wealthy people and poor people tend to concentrate in different areas — a form of social segregation. There are a number of reasons for this.

- **Housing** — wealthier groups can choose where they live, paying premium prices for houses well away from poor areas, with pleasing environments and services such as quality schools and parks. The poorer groups have no choice and have to live where they are placed in social housing, or where they can find a cheap place to rent.
- **Changing environments** — housing neighbourhoods change over time. Many Georgian and Victorian houses have been converted into multi-let apartments for private renting to people on low incomes. Conversely, former poor areas are being taken over by more wealthy people and are being upgraded (gentrification). The right-to-buy legislation of the 1980s transformed many council estates, as houses were bought by their occupants and improved.
- **The ethnic dimension** — newly arrived migrants, being generally poorer, tend to concentrate in poor areas in a city, often clustered into multicultural areas. Such ethnic groupings often persist into later generations.

Examiner tip

Such inequalities exist in most towns and cities — think of areas in a town/city you know well that illustrate them.

The causes of inner-city decline

Economic decline

Since the 1950s there has been a widespread movement of employment away from the large conurbations to smaller urban areas and to rural areas. De-industrialisation in the inner cities was accompanied by the expansion of both manufacturing and service-sector employment in rural areas and small towns. This shift can be partly explained by:

- the changing levels of technology and space requirements of manufacturing industry, which resulted in a shortage of suitable land and premises in the inner cities
- globalisation of production, which led to declining profits and increased competition. To remain competitive, companies were forced to restructure their production methods and labour requirements, which involved movement of investment to new locations in the UK

Employment losses were skewed towards the inner cities as many of the types of workplace most likely to be closed were found here.

Population loss and social decline

During the late twentieth century, the UK's largest conurbations lost 35% of their population and migration was the key cause. Migrants from the inner-city areas have tended to be the younger, the more affluent and the more skilled. This has meant that those left behind are the old, the less skilled and the poor. Therefore, economic decline of these areas has led to social decline.

The poor physical environment

The physical environment of the inner cities is usually poor, with low-quality housing (both Victorian terraced and high-rise flats of the 1960s and 1970s), empty and derelict properties, vacant factories and unsightly, overgrown wasteland. Urban motorways, with flyovers, underpasses and networks of pedestrian walkways, contribute further to the bleak, concrete-dominated landscape.

Political problems

There has been concern that the problems of inner-city residents have been marginalised politically. Inner cities have the lowest election turnouts in the UK, reflecting the degree to which the people feel rejected. In looking for solutions local people have often elected members of far-right parties to local councils.

Examiner tip
You should study the characteristics of, and reasons for, decline in one urban area.

Perception issues

Residents of many urban areas perceive that inner cities have other problems, which are not always easy to deal with. Such issues include 'red light' districts, and other forms of illegality; the areas become forgotten, lack social and economic investment and have the reputation of being areas to avoid.

Regeneration

(See also gentrification, and property-led regeneration above.)

Partnership schemes

City Challenge Partnerships represented a major switch of funding mechanisms towards competitive bidding. To gain funding a local authority had to come up with an imaginative project and form a partnership in its local inner-city area with the private sector and the local communities. The partnership then submitted a 5-year plan to central government in competition with other inner-city areas. The most successful schemes combined social aims with economic and environmental outcomes.

Knowledge check 34
Write down the names of your case study of each of gentrification, property-led regeneration and partnership. The distinction must be clear in your mind.

The City Challenge initiative was designed to address some of the weaknesses of the earlier regeneration schemes. The participating organisations — the partners — were better coordinated and more involved. This particularly applied to the residents of the area and the local authority. Separate schemes and initiatives operating in the same area, as had happened before, were not allowed — the various strands of the projects had to work together. Cooperation between local authorities and private and public groups, some of which were voluntary, was prioritised.

You should study one case study of a partnership, preferably in the UK. You may either examine a partnership under the City Challenge scheme, or you may study one from more recent times. Most redevelopment schemes undertaken within the twenty-first century have been partnership schemes.

Examiner tip
Make sure you reflect on whether the scheme has been successful or not, and provide evidence to support that view.

Retailing and other services

The decentralisation of retailing

The traditional pattern of retailing is based on two key factors:

- easy, local access to goods which are purchased on a regular basis, often daily and particularly so if perishable
- willingness to travel to a shopping centre for goods with a higher value which are purchased less often, such as household and electrical goods

For many years, these factors led to a two-tier structure of retailing. Local needs were met by corner shops in areas of terraced housing, and by suburban shopping parades. Higher-value goods were purchased in the town centre (the central business district or CBD) and required a trip by bus or car. In the last 30 years technology has had a major influence on the patterns of retailing:

- supermarkets, superstores and hypermarkets have been built in residential areas and town centres
- non-food retail parks have also expanded, especially on the outskirts of towns, with easy access to main roads
- large out-of-town shopping centres have been built on the periphery of large urban areas and close to major motorways, often with their own motorway junctions
- in the twenty-first century there is an increase in e-commerce — electronic home shopping using the internet and digital and cable television systems

Examiner tip
It is always better to illustrate such points by referring to examples — many of which may be local to you. This is good practice.

Conversely, in some areas, there are trends towards more traditional farmers' markets which sell local fresh produce for those customers who are willing to pay more than supermarket prices for 'healthy' foods. Equally, in many town centres, specialist retailers (of, for example, vinyl records) are becoming more common.

Knowledge check 35
Why have many retailing areas been established out of town centres?

Factors affecting retail change

Increased mobility

Nearly all the changes arise from increased ownership and use of the private car. Car parking in city centres is expensive and restricted. Out-of-town retail areas have large areas of free car parking. Locations next to motorway junctions offer speedy access.

The changing nature of shopping habits

People now purchase many items as part of a weekly, fortnightly or even monthly shop. The use of freezers in most homes means items that once had to be purchased regularly for freshness can now be bought in bulk and stored. Retailers have also developed more 'ready-made meal' products that can be stored in a domestic freezer.

Changing expectations of shopping habits

An increasing number of people use shopping as a family social activity, involving more than just the act of shopping. Consequently, many of the larger shopping areas combine retailers with cinemas, restaurants, fast-food outlets and entertainment areas.

Out-of-town-centre retailing areas

Knowledge check 36

List three specific characteristics of one out-of-town-centre retailing area you have studied.

Large areas have been devoted to major retail parks involving:

- redevelopment and/or clearance of a large area of cheap farmland or a brownfield site
- the creation of extensive car parks
- the construction of a link to a motorway interchange or outer ring road
- the development of other transport interchange facilities — bus station, supertram, railway station
- the construction of linked entertainment facilities, e.g. Warner Village cinemas, fast-food outlets

Examiner tip

Make sure you have some pieces of information that could only apply to your chosen OOTCR area. This enables a strong 'sense of place'.

You are required to study one out-of-town-centre retailing area in detail.

The central business district (CBD)

The CBD of a city contains the principal commercial areas and major public buildings and is the centre for business and commercial activities. The CBD is accessible from all parts of the urban area and has the highest land values in a city. In some CBDs, retailing is declining because of competition from out-of-town developments. This means there is a greater emphasis on offices and services. Many decision makers are worried that some CBDs are in decline. Dereliction, vacant buildings, increased numbers of low-grade shops (e.g. charity shops and discount stores) and lack of investment all fuel decline and may accelerate the success of out-of-town shopping centres.

However, most CBDs do continue to flourish alongside the new out-of-town locations. A number of strategies have been devised to help maintain the strength of city centres:

- the establishment of management teams to coordinate overall management of CBDs
- the provision of a more attractive shopping environment with pedestrianisation, new street furniture, floral displays, paving and landscaping
- the construction of all-weather shopping malls that are air-conditioned
- the encouragement of specialist areas, cultural quarters and arcades
- the improvement of public transport links to the heart of the CBD
- the extensive use of CCTV and emergency alarm systems to reduce crime
- the organisation of special shopping events such as Christmas fairs and late-night shopping
- conservation schemes, such as the refurbishment of historic buildings in heritage cities

Knowledge check 37

List three specific characteristics of one urban centre (CBD) retailing area you have studied where there has been significant redevelopment.

Many cities are also encouraging the development of functions other than retailing to increase the attractions of a CBD, including:

- encouraging a wider range of leisure facilities, including bars, restaurants, music venues, cinemas and theatres that people visit in the evening — the so-called 'night-time economy'
- promoting street entertainment; developing nightlife and themed areas
- constructing new offices, apartments, hotels and conference centres
- encouraging residential activities to return to city centres

You are required to know a case study of the redevelopment of one urban centre (CBD).

Examiner tip
Make sure you have some pieces of information that could only apply to your chosen CBD area. This enables a strong 'sense of place'.

Contemporary sustainability issues

You are required to study two issues relating to sustainability in urban areas: waste management and transport management. Examine each of these issues in a general sense, but also with good use of a range of case studies. The case studies can come from anywhere in the world — just make sure they are 'fit for purpose' and illustrate well the point you are making.

Examiner tip
'Fit for purpose' means that the case studies are the right ones for the process you are writing about.

Waste management

The average person in the UK produces 517 kg of household waste every year. Not only is the amount of waste increasing, but so is its potential toxicity and the length of time it is toxic for. Waste disposal in the UK is efficient, so people are not generally aware of the problems that waste is creating.

You should be able to discuss a range of strategies that are being used to dispose of waste:

Examiner tip
The use of examples local to you is perfectly acceptable for any of these strategies.

- **Reduction** — businesses are encouraged to reduce the amount of packaging associated with products. Consumers can play a part by refusing to accept plastic bags, or by opting for products that do not use excessive packaging.
- **Re-use** — some re-use of milk containers, soft drinks bottles and jam jars has been attempted. However, the most successful example of re-use is the sale of 'bags for life'. Some shops charge cash deposits on glass bottles to encourage their return.
- **Recycling** — waste products such as paper, glass, metal cans, plastics and clothes can be recycled if they can be collected economically. However, recycling has costs — start-up costs, transport from collection points to the processing plants, and the hot water and other materials needed for cleaning and processing.
- **Energy recovery** — waste material can be converted into energy. The main method is incineration. However, many old, polluting incinerators have been closed down. Some modern incinerators generate electricity or power neighbourhood heating schemes and are regarded by some to be a sustainable option for waste disposal.
- **Composting** — on a small scale, organic waste (kitchen scraps and garden waste) can be used to make compost to fertilise gardens or farmland. On a bigger scale, 'anaerobic' digestion is an advanced form of composting, taking place in an enclosed reactor. The biogases produced can be used to provide an energy supply and the solid residue can be used as a soil conditioner.

Knowledge check 38
How is waste collected within a recycling context in an area you have studied?

- **Landfill** — waste is dumped in old quarries or hollows, which is convenient and cheap. However, it is unsightly and is a serious threat to groundwater and river quality because toxic chemicals can leach out and contaminate the water. Decaying matter at landfill sites also produces methane gas, which can be explosive and is a greenhouse gas.

Transport and its management

The spread of houses into suburbs and small towns and villages of rural areas, while jobs continue to be concentrated in the central parts of cities, has created surges of morning and evening commuters. These surges take place along roads (private cars and commercial buses) and railway networks. Other traffic flows for shopping, entertainment and other commercial services add to the overall problems of mass transportation.

However, despite mounting problems, government policy in the UK has continued to favour private road transport. Government figures show continued expansion of roads, road traffic and progressively heavier goods vehicles. Alongside this there has been relatively slow growth in railway usage. The number of rail passengers has increased, particularly in the London region, but this has been accompanied by problems of poor and inadequate rolling stock, inadequate track maintenance, escalating fares, accidents and poor travelling experiences.

How and why is urban traffic increasing?

Car ownership is increasing throughout the world. Most will be concentrated in developed countries, and in the urban areas within these. There are several reasons for this growth.

- **A large and growing urban working population** — a high proportion of the employed work in the major urban areas of the country, but live in rural or suburban areas. These people make regular journeys to and from their homes by road and rail. However, many commuter journeys are now between one suburb and another, rather than suburb to town centre. Most public systems were developed for travel from suburb to town centre, not across town, and therefore suburb to suburb journeys have to be undertaken by private car, resulting in congestion of suburban roads.
- **Economic growth** — economic growth in retailing and other consumer services has led to more service vehicles on urban roads. Freight traffic, such as delivery vans, is likely to increase as e-commerce becomes more important in retailing.
- **The growth in urban incomes** — earnings in urban areas are usually higher than in rural areas. In addition, incomes have risen faster than the relative rise in car prices, leading to multiple car ownership in many families.
- **The growth in the number of journeys** — as the number of cars increases, so does the number of journeys that people make in them. Many of these extra journeys are for leisure purposes. There are also more journeys to school by car, as fewer children walk.

Knowledge check 39

Outline problems associated with transport in urban areas.

Possible solutions to urban transport problems

- **Road schemes and restricted access** — in the 1970s and 1980s new orbital, ring and radial routes were thought to be the solution. On a smaller scale, the creation of bus lanes with priority at junctions is an effective way of encouraging public transport use and decreasing car traffic.

- **Road traffic management schemes** — many provincial cities suffer from severe traffic congestion but do not have the option of a new ring road or new arterial routes. Some of the strategies that are being introduced here are:
  - strict on-street parking controls, and expensive multi-storey parking
  - restrictions on vehicular access, for example, pedestrianisation of large areas of the centre
  - one-way systems and traffic calming measures
  - encouraging the use of public transport, for example, park and ride schemes
- **The introduction of new mass transit systems** — mass transit systems have been used to provide low-cost public transport from the suburbs to the city centre. Two recent examples in the UK include the Supertram in Sheffield and the Metrolink in Manchester.

Examiner tip

The use of examples local to you is perfectly acceptable for any of these strategies.

Knowledge check 40

Describe one traffic management system in a town or city you know well.

Summary

World cities

- There is a wide range of city sizes and types around the world. The nature of these cities is closely related to levels of economic development.
- Urban areas are undergoing a range of processes, some of which bring people into the urban area, while others take people away from it.
- Each of these urban processes creates issues that need to be planned and managed effectively.
- The relative strength of these processes is also dependent on the level of economic development of the area affected.
- Many established urban areas contain districts that have declined — socially, economically and environmentally.
- Equally, they also contain districts that have undergone a range of strategies aimed at improvement — social, economic and environmental.
- There have been significant changes in retailing activities in the last few decades.
- These changes have impacted on a variety of locations within the urban area.
- Urban growth has also created issues regarding human activities, such as waste and transport management, within urban areas.

Development and globalisation

Development

Development refers to an improvement in a number of different characteristics of a population. There can be:

- **Economic development** — an increase in a country's level of wealth with more employed in manufacturing and services than agriculture and a greater use of energy per capita.
- **Demographic development** — increases in life expectancy with falling death, infant mortality and birth rates.

Knowledge check 41

Explain the links between economic, demographic and social development

- **Social development** — an improvement in a range of features which increases the quality of life of the population. These include education, literacy, medical facilities, sanitation, housing and personal freedom.
- **Political development** — greater freedom to choose who forms a government.
- **Cultural development** — greater equality for women and better race relations in multicultural societies.

Measuring development

The most widely used measures focus on **economic** development. They are:

- **GDP** — the total value of all goods and services produced by a country in a year expressed in amount per head of population
- **GNP** — GDP plus all net income earned by the country and its population from overseas sources

Other measures take into account the local cost of living and demographic and social indicators. A widely used measure is the **human development index (HDI)** which measures life expectancy, adult literacy and school enrolment and the real GDP per capita (based on the purchasing power of people's incomes or PPP).

The development continuum

Knowledge check 42

As an exercise, you could see how many countries you could name in each category.

Examiner tip

It is important that you understand the meaning of the term 'development continuum', and that you are able to see where many of the world's countries are positioned within it. You should also examine the relationship between the North/South divide and the development continuum.

Early classifications of countries recognised three groups:

- **First/Developed World** — western Europe, North America, Japan, Australia, New Zealand
- **Second World** — state-controlled, centrally planned economies, e.g. the former USSR
- **Third/Developing World** — the countries of Africa, Asia and Latin America

The world is much more complex today and many groups can now be recognised, such as developed countries, developing countries, least developed countries (LDCs), newly industrialising countries (NICs), recently industrialising countries (RICs), centrally planned economies, oil-rich countries.

Countries pass from one category to another and from one condition to another. Transition is gradual and with all the world's countries at differing levels of development (and with different types of development), this patchwork is now referred to as the **development continuum**.

The development gap (the North–South divide)

The Brandt Report (1980) first pointed out the development gap between the richer 'North' countries and those of the poorer 'South'. Brandt gave his name to the line which separated them. In that report, it was pointed out that:

- the North accounted for around **80% of global GDP** but only contained **20% of global population**
- the South accounted for **20% of global GDP** but contained **80% of the population**

Since then, globalisation has changed the economic circumstances across the world with some countries experiencing large levels of growth and there is also an increasing interdependence between different parts of the world. To improve their economic standing, states in the South have attempted to:

- pursue autonomous industrial policies
- change trade rules
- secure higher levels of development assistance
- encourage transworld movements which make the global economy more equitable

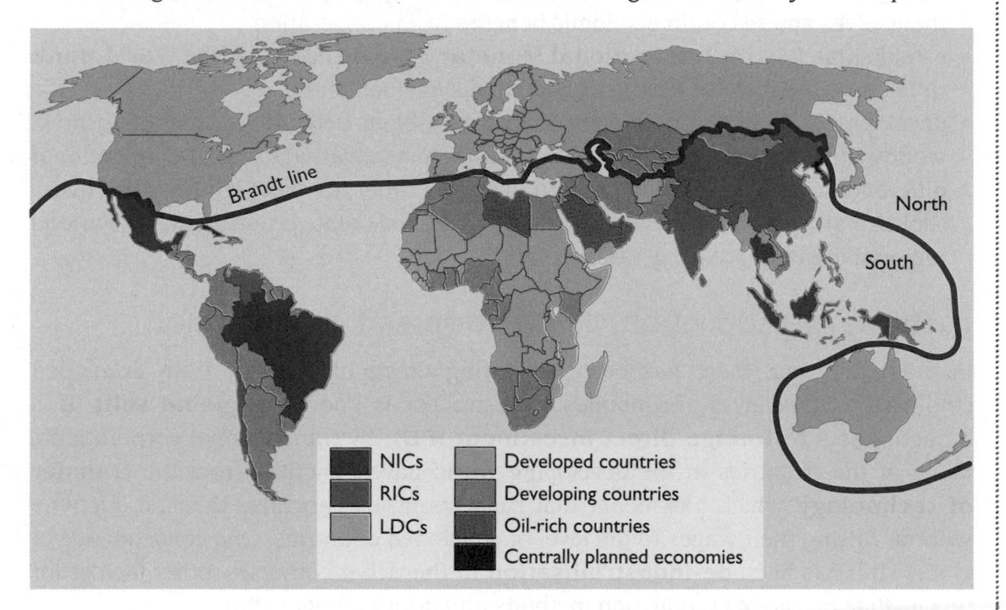

Figure 6 Global economic groupings

Globalisation

Globalisation is the increasing interconnection in the world's economic, cultural, and political systems. On the plus side, globalisation has brought:

- a spread of wealth to more of the world's population
- a greater cultural mix
- access to food, entertainment and music from all over the world

On the negative side, it has brought:

- uncontrolled migration into certain areas
- serious inequalities in wealth
- a heavy cost on the environment such as the effects of global warming

Examiner tip
It is very important that you have a very good understanding of the term 'globalisation' and how it has been brought about.

The progress of globalisation

Many economists believe that globalisation began in the nineteenth century as:

- transport and communication networks expanded rapidly
- world trade grew leading to an increase in interdependence between rich and poorer nations
- capital flows expanded as European companies started operations in other parts of the world — the beginning of transnational corporations

In the later part of the twentieth century, globalisation was shaped by the:

- emergence of free market ideas
- de-regulation of world financial markets

- establishment of the **General Agreements on Tariffs and Trade (GATT)** (later the **World Trade Organization (WTO)**) which sought to gradually lower the barriers to international trade, with free trade as its aim
- emergence of trade blocs whose members sought to stimulate trade between themselves and to obtain economic benefits from cooperation
- establishment of the **International Monetary Fund (IMF)** and the **World Bank** (International Bank for Reconstruction and Development)
- development of global marketing, which has been defined as 'marketing on a worldwide scale, reconciling or taking commercial advantage of global operational differences, similarities and opportunities in order to meet global objectives'. Global marketers view the world as one market and create products that fit various regional marketplaces (e.g. Coca-Cola)

Examiner tip
It is important that you understand what is meant by 'global marketing' and its impact on the economy of various types of countries.

Patterns of production, distribution and consumption

In manufacturing, there has been a filtering down of industry from developed countries to lower wage economies. This process is known as **global shift**. It is brought about by **foreign direct investment** (**FDI**) by transnational corporations. Many of the countries in the developing world have benefited from the **transfer of technology** which has meant that such countries can raise their productivity without raising their wages to the level of developed countries. One consequence of global shift has been **de-industrialisation** in the richer countries (other factors for this include outmoded production methods and poor management).

Provision of **service operations** has become increasingly detached from the production of goods. In the 1990s, a number of transnational service conglomerates emerged, seeking to extend their influence on a global scale, particularly in banking and other financial services, and advertising. A growing trend has been the movement of low-level services, such as call centres, from the developed to the developing world where employment costs are much lower.

Knowledge check 43
How has the global distribution of manufacturing changed in the last 60 years?

Development patterns and processes

Newly industrialising countries

First phase

Since the early 1960s, a number of countries have undergone rapid industrialisation. The initial impetus came from TNCs looking for areas where labour and other costs were considerably lower. Japanese TNCs looked at their less developed neighbours, in particular South Korea, Taiwan, Singapore and Hong Kong (the **'Asian Tigers'**). The advantages of these countries were:

- relatively well educated population with skills
- reasonably well developed communications
- government support through low-interest loans, grants etc.
- less rigid laws and regulations on labour taxation and the environment, allowing for more profitable operations
- strong work ethic in the population

Examiner tip
You should understand how the initial NICs were able to develop and how companies set up by those countries were eventually able to turn themselves into TNCs.

As these economies grew, local entrepreneurs began to set up their own companies e.g. the *chaebol* in South Korea, which in turn became TNCs as they moved operations from the home country.

Second phase

As operating costs began to increase, TNCs began to look for countries which could offer lower wage levels. Countries such as Malaysia and Thailand were a target for this movement of routine manufacturing tasks.

Third phase

Since 1990 both China and India have seen massive TNC investment. The economic growth of China has been both rapid and sustained and is the highest that any country has seen. India's economic growth has been based on the service sector rather than manufacturing and by the early twenty-first century, its service sector accounted for around 50% of GDP.

Growth in the twenty-first century

New markets are now emerging which represent over 70% of the global population. Along with China and India, these include Brazil, Chile, Argentina, Russia, South Africa, Saudi Arabia and the United Arab Emirates (UAE). The momentum for such changes has been the need to:

- raise living standards
- increase opportunities for the population
- attract foreign investment
- raise their markets to international standards

Knowledge check 44

Why were foreign companies so keen to invest in China?

Examiner tip

You need to make a case study of both China and India: China in terms of its rapid economic expansion, and for India, the role of the service sector in the country's growth.

Examiner tip

You need to make a short case study of one of these emerging markets. Dubai is a good example, but make sure that your material is right up to date, including the financial crisis that has affected this country.

Countries at very low levels of economic development (LDCs)

These are the countries described by the United Nations as 'the poorest and most economically weak of the developing countries with formidable economic, institutional and human resource problems, which are often compounded by geographical handicaps and natural and man-made disasters'. Of the 50 countries on the list, 33 are in Africa, south of the Sahara. They are defined by the following:

- low incomes (under $800 GDP per capita over a 3-year period)
- human resource weaknesses, based upon indicators of nutrition, health, education levels and literacy
- economic vulnerability, shown by low levels of economic diversification

Many of them also suffer from widespread conflict (civil wars etc.), lack of political and social stability, and political corruption, often based on governments which are authoritarian.

Knowledge check 45

LDCs are beset by a number of economic and social problems, but they also suffer from other problems which may be the cause of their financial position. Describe some of these problems.

Quality of life

Most of the population has an income too small to meet their basic needs. The resources of the countries, even when equally distributed, are not enough to meet the needs of the population on a sustainable basis. Although there have been attempts

to reduce poverty, the high population growth rate means that the actual numbers living in extreme poverty are increasing. Many of these countries therefore depend upon external financing, with aid payments reaching record levels in the first decade of the twenty-first century.

Debt

Examiner tip

It is very important that you understand how such countries (LDCs) arrived in this situation and the steps that are being taken, by themselves and outside agencies, to improve the situation.

From the 1970s onwards, such countries found themselves in a debt crisis as a result of borrowing money from banks in the developed world. This would make it impossible to reverse socio-economic decline. With little hope of repayment, and interest accumulation which made settlement even more remote, the IMF and World Bank set up the **Heavily Indebted Poor Countries** programme (**HIPC**). This provided debt relief and low-interest loans to reduce external debt repayments to sustainable levels, on condition that the receiving country met a range of economic management and performance targets. Another scheme, the **Multilateral Debt Relief Initiative** (**MDRI**), seeks to cancel much of the debt of such countries.

Social problems

Apart from a lack of income, these countries have a range of other problems. The 2000 UN Millennium Declaration set a number of targets in order to reduce extreme poverty.

Global social and economic groupings

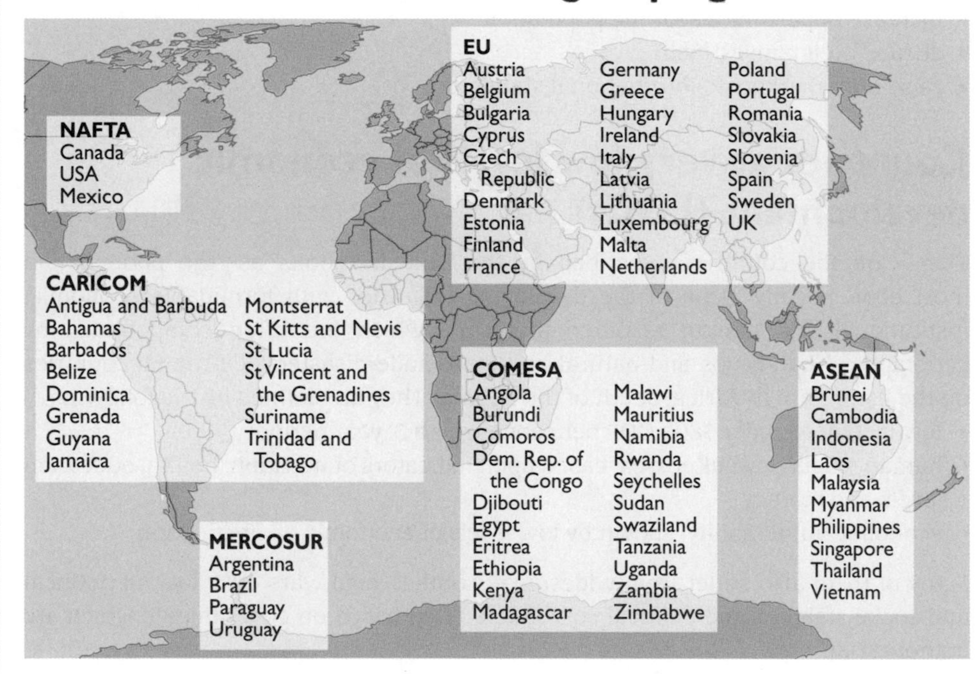

Figure 7 Selected regional trade bloc groupings

Countries seeking to further their economic development have sought alliances to promote trade between countries, and also to provide other benefits. Some of the types of groupings (Figure 7) are:

- **Free trade areas** — tariffs and quotas are reduced or abolished on goods between members; restrictions on goods coming into the area. For example, North American Free Trade Association (NAFTA), European Free Trade Association (EFTA).
- **Customs unions** — member countries operate a tariff on imports from outside the group. For example, Mercosur (Latin America).
- **Common markets** — like customs unions but with greater freedom of movement of labour and capital. For example, the European Union (EU) was previously in this form.
- **Economic unions** — all of the above, but members are also required to adopt common policies in areas such as agriculture, fisheries, transport, pollution, industry, energy and regional development. For example, the current form of the EU.

The consequences of international groupings

Positive:

- greater chance of peace between member nations
- as trade barriers are removed, economies should prosper, giving higher living standards
- particular sectors of a national economy can be supported, e.g. agriculture
- remote regions can receive support from a central organisation, e.g. EU Regional Fund
- people seeking work can move between countries
- possibility of developing a common currency, e.g. the euro
- greater overall democratic function

Negative:

- loss of sovereignty, with centralised decisions
- loss of some financial controls to a central authority such as a bank, e.g. European Central Bank
- pressure to adopt centralised decisions, e.g. in Europe, the Social Chapter, working-hours directive, food regulations
- having to share resources may damage economic sectors, e.g. UK sharing traditional fishing grounds
- elites within a system can hold a disproportionate amount of power through the voting system
- the drive towards federalism is opposed by many
- smaller regions within large countries demand a greater voice which has led to separatist movements

The growth of the European Union

The European Union began life as the European Economic Community (EEC) after the signing of the Treaty of Rome in 1957. Membership has grown to 27 members in 2011.

The aims of economic integration have been promoted through:

- reducing tariffs and other barriers between members
- establishing a common external tariff on imported goods from outside the EU
- allowing free movement of capital and labour
- establishing common policies on agriculture, fishing, industry, energy and transport

Examiner tip

The European Union is named in the specification, therefore examiners are entitled to set questions on the topic. They will assume that you have studied the reasons behind its foundation, its growth, and some of the recent developments within the Union.

There is no doubt that living standards have risen across Europe as a result of membership of the EU. Trade has promoted competition and led to greater efficiency. There has also been the development of a common currency (the euro), which has not been adopted by all members. In addition to the above criticisms, there is also a perception that many seem to be working towards a federal Europe, taking even more power away from national bodies.

A new treaty, the Treaty of Lisbon, was signed in December 2007. Its major features were:

Knowledge check 46

Describe the growth of the European Union.

- creation of a European president
- creation of a common foreign policy
- commissioners to be reduced from 27 to 18
- national vetoes removed in around 50 policy areas
- voting weights redistributed between members

Transnational corporations (TNCs)

A transnational corporation is a company which operates in at least two countries. The organisation is hierarchical, with the headquarters and research and development (R&D) often located in the country of origin, with centres of production overseas. As an organisation becomes more global, regional R&D and even regional headquarters will develop. TNCs take many different forms and include a wide range of companies. They are involved in the following:

- **Resource extraction** — mining, gas extraction and oil production. Many large TNCs operate in this field, such as ExxonMobil, Royal Dutch Shell and BP.

Figure 8 The Mini: an example of an assembly industry

- **Manufacturing** — this takes several forms:
  - **high tech** — computers, microelectronics, pharmaceuticals (Hewlett Packard, GlaxoSmithKline, AstraZeneca)

  - **consumer goods** — motor vehicles, televisions and other electrical goods (Ford, General Motors, BMW, Sony). Many of these are assembly industries (see Figure 8)
  - **mass-produced consumer goods** — cigarettes, drinks, breakfast cereals, cosmetics, toiletries (Coca-Cola, Kelloggs, Unilever)
- **Service operations** — banking/insurance, advertising, freight transport, hotel chains, fast food outlets, retailers such as Citigroup, Barclays, AXA, Allianz, McDonald's, InterContinental Hotels Group (Holiday Inn) and Tesco.

Knowledge check 47

How important globally is the service sector in foreign direct investment?

Growth of TNCs

Companies expanded from their home base to become TNCs for some of the following reasons:

- to take advantage of spatial differences in the factors of production at a global scale. One reason is to look for **cheaper labour costs**
- to take advantage of government policies such as **lower taxes**, **subsidies** and **grants**
- to take advantage of **less stringent legislation** on employment and pollution
- to get round **trade barriers**
- to locate in **markets** where they want to sell
- to grow to a size where they achieve **economies of scale**, allowing them to reduce costs, finance new investment and compete in global markets
- to acquire **geographical flexibility** so that they can shift resources and production between locations at a global scale in order to maximise profit

To serve a global market, TNCs may globalise their production in the following ways:

- produce for the market in which a plant is situated
- use one plant to produce for a number of countries
- use integrated production, where each plant performs part of a process
- source parts in places where they assemble their products close to market, this is known as **glocalisation**

Social, economic and environmental impacts of TNCs

Impacts on the host country

The **advantages** of TNCs locating in a country are:

- **Employment**.
- **Injection of capital into the local economy** — more disposable income will create a demand for more housing, transport and local services.
- **Multiplier effects** — investment by a TNC can trigger more employment through the process of cumulative causation bringing greater wealth into a region, e.g. component suppliers and distributors.
- **New working methods** — the transfer of technology will create a more skilled workforce. Also, new methods will be adopted, such as just-in-time (JIT) component supply and quality management systems.

The **disadvantages** are:

- **Competition** — arrival of TNCs may have an adverse effect on local companies which might not be as efficient.
- **Environmental concerns** — many developing countries have less stringent pollution laws than in the TNC home country.

- **Labour exploitation** — many have alleged that some TNCs exploit cheap, flexible, non-unionised labour forces in developing countries. This has been strongly denied by many TNCs who point to a basic standard of operation involving worker training facilities, and promotion opportunities for locals with a minimum wage in force.
- **Urbanisation** — establishing factories in major urban centres leads to their expansion as younger workers migrate from rural areas. This can also have serious consequences in those rural areas.
- **Removal of capital** — to the TNC's home country.
- **Outside decision making** — plans affecting plants in developing countries are made in the home country and usually for the benefit of the TNC and its profitability.

Impacts on the country of origin

Examiner tip

It is important that you have an understanding of all four areas of impact, i.e. positive and negative impacts on the host country and positive and negative impacts on the country of origin.

The **positive** impacts are:

- **high-salary employment** — even when TNCs move their operations overseas, the headquarters and R&D often stay in the home country
- **return of profits** — successful TNCs return their profits to the home country to be distributed among shareholders. Profits are also taxed which increases government revenues

Negative impacts are:

- **unemployment** — for both the TNC's employees and those in component suppliers
- **reverse multiplier effects** — as unemployment increases in a region, disposable income falls leading to a downward spiral (vicious circle)

Examiner tip

A case study of one TNC should be undertaken. Here Tesco is given as an example.

Case Study

Tesco: a service transnational

One of the world's largest retailers is the UK-based Tesco which has its headquarters in England. The company operates stores in 12 countries outside the UK and in 2009 this totalled over 2,000 stores. In the UK Tesco operates stores of different sizes and product ranges (Tesco Extra, Tesco Metro, Tesco Express, One Stop, Tesco Homeplus). In February 2008 there were over 2,100 outlets in the UK.

Tesco began life in 1919 as a grocery stall in the east end of London, run by Jack Cohen. The first store was opened in 1929 and by 1947 the company had grown to the point where it was floated on the London Stock Exchange. In the second half of the twentieth century, Tesco grew both by opening new stores and by takeovers of existing companies. Originally selling only groceries, the company has expanded into many different areas today. Apart from food and drink, these include:

- clothing
- electrical goods
- retailing and renting DVDs
- books
- CDs and music downloads
- internet services and mobile phones
- insurance (home/car/personal)
- financial services
- petrol
- optical services
- travel

Knowledge check 48

Make a study of another TNC other than Tesco. Make this a company involved in manufacturing to see whether or not it operates in a different way from a service TNC such as Tesco.

The company has increasingly seen other countries not just as places to source goods from, but as markets in which to operate. After opening stores in Hungary and Poland, entry into the Asian markets began in 1998 with a joint venture in South Korea. In 2004, the company entered the Chinese market where rising wealth meant an increasing number of affluent customers. Tesco now employs around over 450,000 people worldwide and operates in the following countries:

- UK
- China (70 stores in 2009)
- Czech Republic (113)
- Hungary (149)
- Ireland (116)
- Japan (135)
- Malaysia (29)
- Poland (319)
- Slovakia (70)
- South Korea (242)
- Thailand (571)
- Turkey (96)
- USA (115)

Recently, the company has begun operations in India, where it is not allowed to open its own-brand retail stores but can sign up to joint ventures and operate wholesale cash-and-carry businesses. In the year 2008/2009, the company had sales of £59.4 billion (inc. VAT) which gave an overall profit of £2,954 million.

Development issues

Trade vs aid

Trade has been seen as a means to allow more revenue to flow into a country, promoting increased wealth and living standards. This is partly dependent on three factors:

- the adoption of Western-style capitalism
- economic growth 'trickling down', providing extra money and resources for new industry to be established
- promotion of free trade, where markets are as open as possible

Some economists doubt that poorer countries with a multitude of problems such as HIV/AIDS, war and drought can ever become developed through trade and economic growth. These arguments are based upon the following:

- LDCs cannot be competitive in world markets because of the great difference in wealth between them and developed countries.
- Many poorer countries depend upon agricultural exports, the price of which has been falling. Also, farmers in richer countries are often protected through schemes such as the EU Common Agricultural Policy.
- Wealth generated by trade does not always 'trickle down' to the majority of the population and the gap between rich and poor in developing countries is growing.
- The debts of many poorer countries have put them in a difficult position. To receive help, they have had to accept suggestions from bodies such as the IMF and World Bank which has often meant cuts in public spending, particularly on health and education.

Examiner tip

You are required to study three development issues: trade versus aid, economic sustainability versus environmental sustainability, and sustainable tourism: myth or reality? Each of these should be studied with reference to contrasting areas of the world.

Aid can be supplied through one of three systems:

- **bilateral** — one government gives to another
- **multilateral** — governments give to international organisations (World Bank, UNESCO), which in turn give to poorer countries
- **non-governmental organisations** (**NGOs**) — many of these are charities, such as Oxfam, which raise money and distribute it to the people who need it most

There are several ways in which aid can be delivered. It does not have to be in the form of money, it could be goods or technical assistance. Distribution can be as:

- **short-term aid** — usually following an emergency (hurricane, flood, tsunami)

- **long-term development projects**
- **top-down aid** — the operation is directed by a responsible body from above, such as in large-scale irrigation and HEP projects
- **bottom-up schemes** — these are 'grassroots' initiatives often funded by NGOs working closely with local people

Critics of aid as a means towards development point out that:
- aid does not always reach those who need it most and is not always used effectively because of corruption
- with a lack of basic infrastructure, it is often difficult to use aid effectively
- aid dependency can be created when aid becomes a substantial proportion of national income
- aid can often come with strings attached, such as having to spend the money on goods from the donor country

Knowledge check 49

For the systems of aid listed above and ways in which aid can be delivered, try to find at least one good example of each feature.

Economic vs environmental sustainability

Sustainability has been defined as 'development that meets the needs of the present without compromising the ability of future generations to meet their own needs'. Within the process:
- human potential (level of wellbeing) is improved
- the environment (resource base) is used and managed to supply people on a long-term basis
- this implies social justice as well as long-term environmental sustainability
- the capacity of the environment to provide resources and absorb increasing levels of pollution is the critical threshold controlling how far population can increase and economies expand

Examiner tip

It is important to understand the question of whether economic sustainability can be compatible with environmental sustainability.

The UN Rio Earth Summit in 1992 set out the sustainability principles listed below.

Environmental principles:
- people should be at the centre of concerns
- states have a right to exploit their own environment, but they should not damage that of others
- environmental protection is integral to the development process
- people should be informed and states should inform them
- there should be environmental legislation and standards within states
- laws should be enacted regarding liability for pollution
- the relocation and transfer of activities and substances that are harmful to health should be prevented
- environmental impact assessments should be carried out for all major economic activities
- states should pass on information on anything which might affect their neighbours

Economic principles:
- the right to development must be fulfilled so as to meet equitably development and environmental needs of present and future generations
- all states should cooperate in eradicating poverty in order to decrease disparities in living standards
- the special needs of the poorest countries should be given priority

- states should cooperate to restore the Earth's ecosystem
- unsuitable production and consumption patterns should be eliminated
- appropriate demographic policies should be promoted
- states should promote an open economic system where trade policies do not contain unjustifiable discrimination
- the internationalism of environmental costs should be promoted with the principle that the polluter should pay

Action on sustainability, to be effective, must involve cooperation between the three 'sustainability pillars': environment, society and the economy.

Knowledge check 50

What is the difference between economic and environmental sustainability?

Sustainable tourism: myth or reality?

The relationship between tourism and the environment has often followed this pattern:

- the environment attracts tourists for scenery or historical heritage
- this should be mutually beneficial as tourism provides revenue to maintain the quality of the scenery and historical heritage
- as tourist flows increase, they can cause major problems, the cost often outweighing the benefits
- this is particularly so when small areas that are vulnerable to damage are involved

Sustainable tourism seeks not to destroy what it sets out to explore. It tries to make sure that:

- it operates within the capacity for regeneration and future productivity of natural resources
- the contributions of local people and their culture are recognised
- there is an acceptance that local people must have an equitable share in the economic benefits of tourism
- everything is guided by the wishes of the local people and communities

Following the Rio Earth Summit, an environmental checklist was drawn up naming areas in which travel and tourism operations could take action. These included:

- waste minimisation, reuse and recycling
- energy efficiency, conservation and management
- management of freshwater resources
- waste-water management
- transport
- land-use planning and management
- involvement of staff, customers and communities in environmental issues

Examiner tip

Make sure that you have in your own mind a clear definition of the term 'sustainability'. Then apply this to the tourist industry.

Ecotourism is one of the fastest growing sectors of the tourism industry. It claims that it is conserving the environment for future generations by planning and management of tourism environments. It has been developed largely by small, dedicated tour companies, but critics have maintained that many of the operations are ordinary tourism dressed up as politically correct, and have even gone as far as to describe them as 'egotourism'. Larger companies, particularly hotel chains, are seeking to become more environmentally aware, but cynics suggest that such strategies are designed largely to generate a good public relations image.

Examiner tip

It is important to understand the principles of sustainable tourism and whether it can ever be a widespread reality.

Summary

Development and globalisation

- Development refers to an improvement in a number of different characteristics of a country's population (economic, demographic, social, political and cultural). It can be measured in a number of different ways. Transition from stage to stage is gradual with all countries at different levels of development; this is known as the development continuum.
- Globalisation is the increasing interconnection in the world's economic, cultural and political systems. This has brought about changes to the global patterns of production, distribution and consumption.
- Since the early 1960s, many countries have developed with increasing industrialisation. Early impetus came through TNCs, particularly in southeast Asia and China. India's development has been through a massive increase in the service sector. In recent years new markets have emerged which have attracted much inward investment.
- There are a number of countries that are still economically weak with formidable economic and human resource problems (least developed countries). A large majority of these LDCs lie in Africa, south of the Sahara. Some efforts are being made by the international community in an attempt to bring about some solutions to these problems.
- Countries seeking to further their economic development have joined together to promote trade and provide other benefits. All groupings bring benefits to their members but there are also negative aspects of such alliances. One of the best known of these alliances is the European Union which began life in 1957 with the signing of the Treaty of Rome.
- One of the driving forces of globalisation has been the growth of transnational corporations (TNCs). These hierarchical companies take many different forms, being involved in resource extraction, manufacturing and increasingly in service operations. Their spread brings advantages and some negative impacts to both the host and donor (origin) countries.
- There are three development issues which you are required to study with reference to contrasting areas of the world. They are:
  – trade vs aid, particularly in the development of poorer countries
  – economic vs environmental sustainability
  – sustainable tourism: myth or reality? Reference must be made here to ecotourism.

Contemporary conflicts and challenges

The geographical basis of conflict

You should be aware of the meaning of a variety of terms in connection with this option, and be able to apply them to a geographical context. Table 10 (on p. 64) is an example of the type of activity you could do.

- **Challenge** — a task or issue that may be perceived as being provocative, threatening, stimulating or an incitement to debate.
- **Conflict** — a state of discord or disagreement caused by the actual or perceived opposition of needs, values and interests between people. In a geographical sense it is often the result of opposing views over the ways in which a resource might be developed, or used.

Causes of conflict

Identity is a sense of belonging to a group or area where there is the same generic character, or a similarity of distinguishing character or personality. Identity can be evident at a number of scales: **national** — loyalty and devotion to a nation; **regional** — consciousness of, and loyalty to, a distinct region within a homogeneous population or nation; **local** — an affection or partiality for a particular place.

Ethnicity refers to the grouping of people according to their ethnic origins or characteristics. More recently the term has broadened in meaning to refer to groups of people classed according to one or more of common racial, national, tribal, religious, linguistic, or cultural origins or backgrounds.

In a geographical sense, **culture** refers to the customary beliefs, social norms and traits of a racial, religious, or social group and to the set of shared attitudes, values and practices that characterise that group.

Territory refers to a geographic area belonging to, or under the jurisdiction of, a governmental authority. The territory may be an administrative subdivision of a country, or a geographic area (such as a colonial possession) dependent on an external government but having some degree of autonomy.

Examiner tip

Think of areas in the world, at any scale, where conflict has been caused by any one, or a combination, of these factors.

Ideology is a systematic body of concepts regarding human life or culture. It can result in a set of integrated assertions, theories and aims that together constitute a **socio-political** programme. Some ideologies can be extreme and at odds with those elsewhere in the world and their supporters may seek to impress their views on others by force.

Patterns of conflict

There are four main scales of conflict in the world:

- **international** — where conflict involves the participation of more than one country
- **national** — where the conflict takes place within a country
- **regional** — where conflict takes place within an area of one country, or across the borders of one or more countries
- **local** — where the conflict is restricted to a small part of one region of a country

Knowledge check 51

Distinguish between nationalism and regionalism.

The expression of conflict

Non-violent conflict does not involve any force or armed struggle. Statements of discontent are offered by word, sign, marching or by silent protest.

Political activity relates to groups operating within a country who seek to acquire and exert political power through government. The groups, known as parties, develop a political programme that defines their ideology and sets out the agenda they would pursue should they win elective office or gain power through democratic means. Political activity often involves **debate** — the formal discussion of a motion before a deliberative body according to the rules of parliamentary procedure.

Terrorism refers to the systematic use of fear as a means of coercion to a political, or more frequently an ideological, end.

Knowledge check 52

Distinguish between terrorism and insurrection.

Insurrection is an act or instance of revolt against civil authority or an established government, usually involving rebellion against the rules of that government.

War is a state of open and declared armed hostile conflict between states or nations. The armed forces of the states involved are the main protagonists of the conflict.

Conflict resolution concerns the means by which conflict at a variety of scales can be brought to an end. In some ways, the expression of conflict as given above — debate, political activity, war — can lead to its resolution. Conflict resolution can be undertaken at a range of scales, from local (such as a planning committee) to international (such as the work of the United Nations).

Table 10 Examples of types of conflict

Type of conflict	Location	Conflict
Civil war	Rwanda	Fought between the Hutu regime and rebel Tutsi exiles with support from Uganda, culminated in 1994 with the mass killing of hundreds of thousands of Tutsis by Hutu militia
	Libya	The dispute between government and anti-Gaddafi forces within Libya (2011)
Separatism	Basque region of France and Spain	A movement advocating either further political autonomy or full independence. Responsible for many assassinations and terrorist acts in Spain (1980s–present)
Insurgency	Iraq and Afghanistan	A diverse mix of militias and foreign fighters using violent measures against the United States-led coalitions and national governments (2003–present)
Invasion	Kuwait	The occupation by Iraqi troops of Kuwait in August 1990. This was met with subsequent expulsion of Iraqi troops in January 1991 by coalition forces
Terrorism	USA	On 11 September 2001, al-Qaeda launched an attack on the World Trade Center by hijacking two aeroplanes, taking over the controls and flying them into the building
	UK	The 7 July bombings of a bus and tube trains in London, 2005
Ethnic cleansing	Former Yugoslavia	Forty-thousand Bosnian Muslims in Srebrenica were targeted for extinction. They were deliberately and methodically killed on the basis of their identity
Border dispute	Kashmir	Parts of this region are now claimed by three countries: India, Pakistan and China. Occasional small-scale acts of aggression to assert sovereignty occur
Civil disobedience	China	The protests of students in 1989 resulted in the massacre of protesters in the streets to the west of Tiananmen Square
	Egypt	The uprising in Tahrir Square, Cairo, in early 2011
Non-violent	Ukraine	In 2004/2005 people protested in the streets of the capital, Kiev, at the outcome of the presidential election. The demonstrations went on for several days and were known as the Orange Revolution
Political activity	UK	The eventual persuasion of various UK governments to grant devolved power to the Welsh and Scottish parliaments

Conflict over the use of a local resource

You are required to study one conflict over the use of a local resource (e.g. land, buildings, space) that is taking place, or has taken place in recent years. Your study should cover:

- *the reason for the conflict, and the attitudes of different groups of people (or participants) to the conflict*
- *the processes (market processes or planning processes) which operate to resolve the conflict*
- *recognition that some people benefit, whereas others may lose, when the outcome is decided*

Examiner tip

It is better to study a conflict that either has ended, or has been running for a few years. This allows you to cover all aspects of potential questions.

There have been a number of conflicts in the UK over large building projects that have achieved public and national notoriety. These include the Newbury bypass, the second runway at Manchester airport, the new container terminal in the Southampton area, and Terminal 5 at Heathrow airport. There are also many examples locally across the UK where people have disagreed about the construction of a new shopping complex, landfill site, housing estate, wind farm. You could investigate a similar conflict in another part of the world.

Examiner tip
Make sure you know some specific facts that can only apply to your chosen conflict.

Such conflicts are resolved by market processes, planning processes or, in some cases, a combination of the two.

Market processes operate where the ability to pay the going rate takes precedence over any local or national concerns. Often, objectors cannot afford to outbid the developer and the development goes ahead with the minimum of consultation.

Knowledge check 53
Which of market processes and planning processes is more dominant in the UK? How will the new 'Localism Bill' fit into these processes?

Planning processes attempt to provide a means by which local authority planners:

- listen to the local community (more democratic)
- listen to the organisation responsible for a proposal
- have overall development control

A local authority's refusal to grant planning permission may lead to an appeal by the developer, either to the local planning committee or to a higher body, for example the Department for Environment, Food and Rural Affairs (DEFRA).

The geographical impact of international conflicts

Knowledge check 54
What is meant by the term 'international conflict'?

The specification requires the study of the 'social, economic and environmental issues associated with major international conflicts that have taken place within the last 30 years' through the examination of one or more case studies. A number of potential areas for study exist in the world, some arising from factors that have emerged in recent years, others have a more historical background.

Examiner tip
When studying your chosen conflict ensure that you concentrate on each of the social, economic and environmental issues associated with it. In many cases these issues are interrelated but, equally, you must be able to separate them out for an examination question.

Some international conflicts you could study are:

- various aspects of the middle east: the West Bank of the River Jordan in Israel/ Palestine, the Gaza Strip, Beirut
- the civil unrest in the Darfur region in Sudan, which became international because of refugees escaping over borders and the deployment of international peacekeepers. In July 2011 the new state of South Sudan was created
- the East Timor conflict during the early part of this century
- Afghanistan
- recent events in north Africa and the Arab world — again these have an international dimension due to refugees and aspects of conflict resolution

Multicultural societies in the UK

Migration of ethnic groups leads to the creation of multicultural societies. In most countries there is at least one minority group and, while they may be able to live peacefully with the majority, it is more likely that there will be a certain amount of prejudice and discrimination leading to tensions and conflict. Multicultural societies are often the product of migration, but they may also be the stimulus for it, as persecuted groups seek to escape oppression. The level of integration of minorities

Knowledge check 55
What is a multicultural society?

varies between societies. In some societies there is a lack of integration, whereas in others there are more tolerant attitudes.

Ethnic segregation is the clustering together of people with similar ethnic or cultural characteristics into separate urban residential areas. The largest ethnic minority in the UK is the Indian population, which forms 27% of the total ethnic minority population. The next largest is the Pakistani ethnic minority (17%), followed by the black Caribbean (15%). Smaller, but still significant, ethnic minorities of Bangladeshi, black African and Chinese people also live in the country. In addition to these 'ethnic' minorities, there are also significant numbers of migrants from other parts of the world, particularly Germany, the USA and most recently Poland. Migration into the UK is so great that migrants are now classed as 'born abroad'.

Examiner tip
The results of the 2011 census will be published by the Office for National Statistics. You can visit its website at: **www.statistics.gov.uk/hub/population/index.html**

In the 2001 census, there were 57 million people in the UK. Of these, 4.3 million (7.5% of the population) were people born abroad. This had increased to 5.5 million in 2008. There is no doubt that immigration will continue, though the areas of origin are likely to change.

The geographical distribution of cultural groups

Ethnic minorities are concentrated in the major urban areas of the country, particularly London (1.8 million migrants) and the southeast (0.6 million migrants), the west and east midlands, Manchester and West Yorkshire. Over 50% of ethnic minorities live in London and the southeast, which has only 30% of the white population, so the concentration here is highest. A significant proportion of ethnic minorities consists of people born in the UK, descended from migrants who arrived from the former Commonwealth countries in the 1950s, 1960s and 1970s.

Examiner tip
Patterns of ethnicity using the 2011 census will be available at **www.neighbourhood.statistics.gov.uk**. You should also be aware of new sets of data as they emerge.

Some variations in the geographical distribution of ethnic groups result from factors in the early days of immigration such as employment in specific industries. For example, there are large concentrations of the Indian ethnic minority in the east and west midlands (e.g. Leicester) and Greater Manchester. The Pakistani minority is concentrated in parts of Bradford, Leeds and Birmingham, and there are large Bangladeshi communities in Luton, Oldham and Birmingham.

Issues related to multicultural societies

Housing

As migrants are often a source of cheap labour, working in low-paid construction, transport or health service jobs, they have tended to concentrate in the areas of poorest housing in major cities. Such concentrations are reinforced by later migrants who seek the support and security of living near friends and relatives within an ethnic community. More recently owner occupancy has increased and some more wealthy individuals have moved into suburban areas.

Education

Concentrations of minorities in inner-city areas have led to some schools being dominated by one ethnic group, which has affected education requirements. For example, special English lessons may be needed for children and their parents (mothers in particular), and bilingual reading schemes may be introduced. In some areas, special religious provision for minority groups has developed into separate schooling, known as 'faith schools'. There is greater integration in communities such as Leicester and Bradford where holiday patterns, school timetables and school meals are being modified to reflect the ethnic mix of the areas.

Healthcare

Literature has been produced in ethnic minority languages to inform parents about the benefits of immunisation and other aspects of healthcare. However, this is now less of a concern as the number of second and subsequent generation ethnic minorities increases.

Social

Second-generation migrant children often have different aspirations from their parents. They are more likely to integrate, and this can cause tension within the ethnic group if they adopt the culture of the host country. The issue of arranged marriages still causes some problems.

Religion

Migrants from the Indian subcontinent, and other parts of Asia, follow a different religion from the host population and this may cause friction with employers when migrants wish to adhere to their own religious calendars and practices.

Economic issues

In the UK, there has been legislation on anti-racism, employment rights and equal opportunities to combat discrimination, prejudice and racism. However, the cost of state benefits for migrants' housing, education and unemployment may still cause resentment and racial intolerance from members of the host population. Migrants now account for one in eight of the UK's working-age population, and they have had a beneficial impact on economic output. In 2011 a report stated that migrants from eastern Europe had boosted the UK economy by £5 billion. Much of this labour is employed in unskilled and low-skilled work. Contrary to popular belief, most immigrants are not employed in manual work. Banking and finance employ 13% of migrants, followed by the hotel and restaurant trade (12%). Construction accounts for just 7% while another 5% work in agriculture and fishing.

Knowledge check 56

Identify and describe one benefit associated with multicultural societies in the UK.

The above are examples of the range of issues that have arisen in multicultural societies in the UK. You should keep an eye on the media, and refer to specific instances of such issues, and others, in actual locations. This use of exemplars will raise the quality of your answer to questions on this topic.

Examiner tip

Although there are issues associated with multicultural societies overseas, all your examples must be UK based.

Separatism

The nature of separatism

When the people of a region feel alienated from central government, they often seek to gain more political control, autonomy or independence.

The **reasons** for separatist pressure in a region include:

- an area which is economically depressed compared with a wealthier core
- a minority language or culture with a different history
- a minority religious grouping
- the perception that exploitation of local resources by national government produces little economic gain for the region
- a peripheral location to the economic/political core
- the collapse of the state, weakening the political power that held the regions together (e.g. the former USSR and Yugoslavia)
- the strengthening of supranational bodies such as the EU, which has led many nationalist groups to think they have a better chance of developing economically if they are independent

There are examples of separatist movements all over the world. Some have succeeded in their aims, while other struggles for independence are ongoing. In a few cases they have become bitter and violent. Some of the best known are:

- in Spain, the **Basque** area (northern Spain and southwest France) and **Catalonia** (northeast Spain), which now have the autonomy to decide many of their own affairs
- the collapse of **Yugoslavia** and the formation of Croatia, Slovenia, Bosnia-Herzegovina, Montenegro and FYR Macedonia
- **Kurdistan**, which is a region that comprises southeastern Turkey, northeastern Iraq, northeastern Syria and northwestern Iran, where the ethnic Kurdish population is the majority
- in Canada, where there are demands for independence for French-speaking **Quebec**, and pressure from the **Inuits** in the north
- national groups *within* former Soviet republics seeking independence, for example **Chechnya** in the Russian republic
- the bitter struggle against the Sinhalese majority in Sri Lanka by the **Tamils**, who want to set up their own state in the northern part of the island. It is claimed that this conflict is now at an end
- **Western Sahara**, which has been fighting for independence since 1975 when armed forces occupied the country and incorporated it into Morocco following Spain's withdrawal
- **Scottish** and **Welsh** nationalism. To some extent the drive for independence was satisfied by the establishment in 1999 of a Parliament with limited tax-raising powers in Scotland and the creation of a Welsh Assembly (with some devolution of decision-making powers, but not tax raising)

Examiner tip

Questions on this topic will require detail of one case study (*depth*), but also some understanding of a range of other areas in less detail (*breadth*).

Knowledge check 57

What is meant by the terms 'autonomy' and 'separatism'?

The **consequences** of separatist pressure may be peaceful or non-peaceful. Those desiring various levels of autonomy have used a wide range of activities to create or press for it. In increasing order of extremism, they include:

- the establishment and maintenance of societies and norms with clear separate cultural identities within a country (e.g. the Bretons in France)
- the protection of a language through the media and education (e.g. Welsh, Catalan)
- the growth of separate political parties and devolved power (e.g. the Scottish and Welsh Nationalists)
- terrorist violence (e.g. the Basques, the Kurds, Chechnya)
- civil war (e.g. East Timor, Sri Lanka)

You should study the nature of, reasons for and consequences of separatism through a range of case studies.

Examiner tip
Make sure you keep up to date on this area of study.

The challenge of global poverty

One definition of poverty is derived from the **international poverty line** which is based on a level of consumption in low-income countries. The international poverty line is currently set at $1.25 a day, measured in terms of purchasing-power parity (PPP).

In recent decades there has been increasing concern about the imbalance between population growth and the resource base of the world. In particular, there have been worries about inequalities in economic growth, development and welfare between countries. At the lowest end of the inequality scale lie those people living in poverty.

A number of indicators have been used to measure development, welfare and poverty:

- **economic indicators** — GNP and GDP
- **demographic and social indicators** — birth and death rates, fertility rates, life expectancy, access to drinking water, children enrolled in primary school, adult literacy, number of people per doctor etc.
- the **physical quality of life index** (PQLI)
- the **human development index** (HDI)

Knowledge check 58
How can poverty be measured?

Using these indicators, the United Nations has assessed the progress made in reducing global poverty through the Human Development Report (HDR). It has noted that:

- in the past 50 years, poverty has fallen more than in the previous 500 years
- poverty has been reduced in some respects in almost all countries
- death rates of children in the developing world have been cut by more than half since 1960
- malnutrition has declined by almost one-third since 1960
- since 1960, the proportion of children not in primary education has fallen from more than one-half to less than one-quarter

On the negative side, the HDR points out that there are still substantial problems, including:

- one-fifth of all people in the developing world still live in poverty, with nearly 1 billion living below the international poverty line
- nearly 1 billion people are illiterate; one child in five does not complete primary school
- some 840 million people go hungry or face food insecurity
- more than 1.5 billion people lack access to safe drinking water
- women are disproportionately poor; half a million women in the developing world die in childbirth each year
- in much of the developing world, the HIV/AIDS pandemic continues to spread unchecked. More than 15 million children lost one or both parents to the disease in 2005 and the number of AIDS orphans is expected to double by 2010.

Examiner tip
You can keep up to date by visiting the Human Development Report website at: **http://hdr.undp.org/en/**

The global distribution of poverty

Worldwide the number of people in developing countries living below the international poverty line fell to 1.4 billion in 2005 — down from 1.8 billion in 1990. The proportion of people living in poverty fell to 27% over this period.

Levels of poverty in east, southeast and south Asia fell due to rapid economic development in these areas. In contrast, poverty rates in western Asia trebled between 1990 and 2005. In sub-Saharan Africa, the proportion of people living in poverty fell from 58% in 1990 to 51% in 2005. Most of this progress has been achieved since 2000.

Much of the poverty in the developing world occurs in rural areas. In these areas there are long-term problems of malnourishment made worse by shorter-term disasters. Floods, drought, plagues of locusts and wars take place in many countries at different times and in different years. These add to the endemic problems arising from low economic development.

Addressing poverty on a global scale

The United Nations **Millennium Development Goals** (**MDGs**) were originally developed by the OECD and emerged from the eight chapters of the United Nations Millennium Declaration, signed in September 2000. The UN Millennium Declaration established 2015 as the target date for achieving most of the MDGs, with 1990 generally used as a baseline. The eight goals are aimed at the global *causes* of poverty. They are to:

- Eradicate extreme poverty and hunger.
- Achieve universal primary education.
- Promote gender equality and empower women.
- Reduce child mortality.
- Improve maternal health.
- Combat HIV/AIDS, malaria, and other diseases.
- Ensure environmental sustainability.
- Develop a global partnership for development.

Examiner tip
Keep up to date with progress on the MDGs by visiting the campaign's website at: **www.un.org/millenniumgoals/**

In September 2010 world leaders came together in New York for a 3-day summit to renew their commitment to achieving the MDGs and to set out plans and practical steps for action. The outcome document — *Keeping the Promise: United to Achieve the Millennium Development Goals* — reaffirmed world leaders' commitment to the MDGs and set out an action agenda for achieving the Goals by 2015. It also affirmed that, despite setbacks due to the current economic and financial crises, remarkable progress had been made on fighting poverty, increasing school enrolment and improving health in many countries, and stated that the Goals remained achievable.

In a major push to accelerate progress on women's and children's health, a number of heads of government from developed and developing countries, along with the private sector, foundations and other international organisations pledged over $40 billion in resources over the next 5 years. The *Global Strategy for Women's and Children's Health* — a concerted effort initiated by UN Secretary-General Ban Ki-moon — has the ambition of saving the lives of more than 16 million women and children, preventing

33 million unwanted pregnancies, protecting 120 million children from pneumonia and 88 million children from stunted growth due to malnutrition, advancing the control of deadly diseases such as malaria and HIV/AIDS, and ensuring access for women and children to quality health facilities and skilled health workers. 'We know what works to save women's and children's lives, and we know that women and children are critical to all of the MDGs,' Secretary-General Ban Ki-moon said. '...we are witnessing the kind of leadership we have long needed.'

There has been criticism in the recent past that international support has not been forthcoming as promised. The commitments of aid made in Monterrey in 2002 and the Gleneagles summit in 2005 have not reached the poorest area, sub-Saharan Africa. At the 2005 Gleneagles G8 meeting it was agreed to increase overseas development aid (ODA) to 0.7% of GNP yet, in 2006 and 2007, it declined by 15%. The developed countries commit solemnly to increase ODA and then immediately reduce it. Critics have stressed that not only must donors fulfil their promises but African countries in particular must be given every opportunity to reach the targets. Toward this end, they stress the importance of trade in supporting sustainable development. Trade can bring prosperity and jobs and can give to the states the means to bring basic services to the people. Restrictive practices by developed countries need to be removed to allow freer trade by developing countries.

Knowledge check 59

What is the G8 group of countries?

However, despite these reservations, several countries have made real progress in saving children's lives, including Malawi, Bangladesh, Nepal and Ghana. These countries have challenged the myth that a country's wealth directly relates to how many children's lives it is able to save. Political will seems to be the single most important thing in saving children. Malawi, for example, has a per capita income of less than US$1 a day, but has more than halved child mortality from 22% in 1990 to 10% in 2008.

Think global, act local

Considerable debate takes place between development experts as to how best to raise living standards in the developing world, particularly in remote rural areas where environmental conditions are harsh and constraints are enormous.

Much government and World Bank aid has funded large capital projects (for example, mega-dams that are intended to be multi-purpose catalysts to regional development). In theory, the wealth that is generated by such **top-down** projects trickles down to the poorer peripheral areas. In fact, many of these projects make the lives of the rural poor worse, rather than better, and they have been severely criticised. **Bottom-up**, small-scale projects are better at raising living standards in poor areas. This is because the development is initiated in consultation with local people and is more targeted to local needs.

You should examine at least one 'bottom-up' scheme or project to illustrate how ***appropriate or intermediate technology*** *can help to raise living standards in poor areas of the world.*

Knowledge check 60

What is an NGO?

Examiner tip

The work of NGOs such as CAFOD, OXFAM and Save the Children could be investigated to find excellent examples of such projects.

Knowledge check 61

What is meant by the term appropriate (or intermediate) technology?

Examiner tip
Be prepared to discuss alternative viewpoints with regard to this quotation.

'No development without security and no security without development'

This issue concludes this option. It can be examined at a range of scales and in a variety of contexts. At one level, you could examine this in the context of a country that is undergoing problems of increasing its prosperity, such as Afghanistan or Somalia, due to a lack of political stability. At another level, you could examine the issue in the context of an area such as Bangladesh where farmers lack security of tenure and security against the forces of nature, and hence further development of their lands is difficult. Similarly there are issues of 'water security' in the area of the West Bank of Palestine/Israel. The key idea throughout is that the two processes of increasing development and increasing security must complement each other.

Knowledge check 62

Afghanistan and Somalia have sometimes been described as 'failed states'. What does this mean?

Summary

Contemporary conflicts and challenges

- Conflicts can arise for a wide range of reasons, which may be interrelated.
- Conflicts can occur over a range of scales, and be expressed by the participants in different ways.
- Conflicts can occur at a local scale where people may disagree about the development or site of a new activity.
- Conflicts also occur at an international scale, and these will have major social, economic and environmental impacts on the area(s) affected.
- Multicultural societies are widespread in the UK; they have arisen due to processes that have operated over several decades.
- Multicultural societies present a range of issues for the various groups involved.
- Separatist movements exist in a variety of locations around the world. They exist for a variety of reasons and have multiple consequences.
- Poverty is one of the greatest challenges that the world faces today — there are multiple causes, and many suggestions as to how it can be addressed.
- For people to develop and have better standards of living, they must feel safe in the areas where they live; yet it is the same people that cause uncertainty and instability.

Questions & Answers

In this section of the book two questions on each of the options are given, one complete structured question and one essay question.

Each structured question is worth 25 marks. You should allow 45 minutes to answer each question, dividing the time according to the mark allocation for each part.

Each essay question is worth 40 marks. You should allow 60 minutes to answer each question, including formulating a plan. You should think in terms of writing within a range of 750–1250 words.

The section is structured as follows:

- sample questions in the style of the examination
- mark schemes in the style of the examination
- example candidate answers at a variety of levels
- examiner's commentary on each of the answers (indicated by the icon )

You should read the commentary with the mark schemes to understand why credit has or has not been awarded. For the weaker answers, the commentary highlights areas for improvement, specific problems and common errors such as lack of clarity, weak development, lack of examples, irrelevance, misinterpretation and mistaken meanings of terms. In all cases, actual marks are indicated.

All questions are marked using a 'levels' system, to a maximum of four levels at A2. Study the descriptions of the 'levels' carefully and know the requirements (or 'triggers') necessary to move an answer from one level to the one above it. The essays are marked using the generic mark scheme on pages 6 and 7, though for each essay you are provided with suggestions of the appropriate content, and the ways in which you can achieve synopticity.

Plate tectonics and associated hazards

Question 1

(a) Study Figure 1 which shows the global distribution of earthquakes.

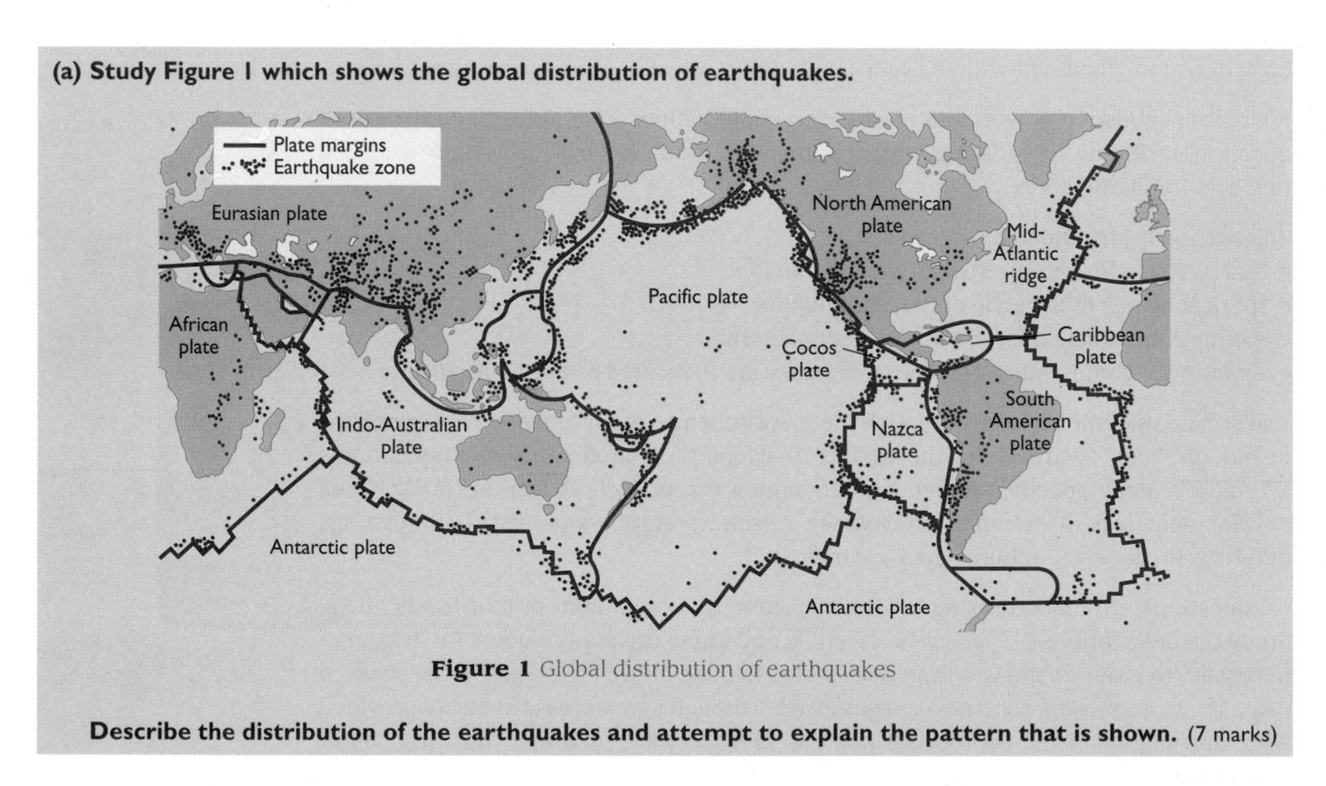

Figure 1 Global distribution of earthquakes

Describe the distribution of the earthquakes and attempt to explain the pattern that is shown. (7 marks)

Note the use of two commands — you have to answer both elements of the question.

Mark scheme

Level 1: basic/simple statements with regard to distribution, mainly names of countries or areas affected by earthquakes. Explanation confined to plate boundaries. (1–4 marks)

Level 2: wider picture seen, such as the 'Pacific Ring of Fire'. Some reference to the large areas where there are no recorded instances. More details on plate boundaries such as the name, the activity taking place there which results in earthquakes and activity within fold mountains. Some attempt to explain why some areas are free/relatively free from earthquakes. (5–7 marks)

Student answer

Earthquakes occur all over the world, but are particularly found in a belt which stretches right around the Pacific Ocean. This belt is associated with widespread volcanic activity and is known as the 'Pacific Ring of Fire'. Earthquakes are mainly found on plate boundaries where the plates are moving relative to each other. When this movement is 'jerky' (sticks for a time, then suddenly releases), earthquakes occur. This happens in California where the North American plate is moving alongside the Pacific plate at what is known as a conservative margin. When plates move apart or meet, with one going under the other, both volcanoes and earthquakes occur. Plates move apart in the centre of the Atlantic Ocean and this is marked on the map by a line of earthquakes along the Mid-Atlantic ridge. In the west of South America, the Nazca plate is subducting under the South American plate causing another area of earthquakes, particularly associated with the fold mountains that are formed there, the Andes. Some areas have very few earthquakes because they are away from plate boundaries, and also where the rocks are very often old and quite stable. Africa and Australia are good examples of such areas, as they are shown to have few earthquakes on the map.

e **7/7 marks awarded.** This is a very good answer. The distribution takes a wider view, but also focuses on specific plate boundaries where earthquakes occur and tries to explain the movement taking place on the particular boundary. Explanation is therefore worked in with facts on earthquake distribution. Specific references are made to California, the Andes and the mid-Atlantic ridge. The answer also makes reference to areas that have very few earthquakes and tries to explain why that is the case. This answer fits the descriptor for Level 2 within the mark scheme and would be awarded the maximum of 7 marks.

(b) Describe the effects that a major earthquake can have on the population of an area. (8 marks)

e Note the word 'effects' — i.e. more than one. You could consider social and economic effects, as well as primary and secondary ones.

Mark scheme

Level 1: simple statements giving nothing more than a list of the effects. (1–4 marks)

Level 2: recognises that effects can be divided into primary and secondary and gives specific examples of each (accept candidate's own definition of primary/secondary as long as it is logical). Links are clearly made between primary and secondary hazards, e.g. ground shaking can cause buildings to fall, breaking gas pipes which could lead to fires breaking out. Several effects can be linked together in a chain; another example could be ground shaking resulting in dams cracking, collapsing and leading to flooding downstream. (5–8 marks)

Student answer

The first, or primary, effect of an earthquake that is felt on the surface is ground shaking. In populated areas this causes widespread damage to buildings. When these buildings are poorly constructed, this can result in almost total destruction as happened in parts of Gujurat (India) in the 2001 earthquake. The soil, when violently shaken, can liquefy also causing buildings to fall down, and landslides and avalanches may rush down slopes which have been moved. The secondary effects are fires, which result from either gas mains breaking or electricity pylons falling over, and disease, after sewage systems are disrupted allowing raw sewage to enter water systems and even appear on the streets. When earthquakes occur near coasts, tsunamis can be created. These are giant waves which travel vast distances over oceans as was shown by the Indian Ocean tsunami in 2004 which drowned thousands of people in countries bordering the ocean. People in poorer countries suffer from food and water shortages after earthquakes occur, the length of time they are without these basic necessities depends on how fast relief supplies can be organised. It may also take a long time for the economy of such areas to recover. Some of these secondary effects therefore can last a long time after the initial ground shaking of the earthquake.

e **7/8 marks awarded.** This answer recognises that there are primary and secondary effects and gives examples of each. Examiners will accept any division into primary and secondary as long as your definitions are logical. (For example, some students will state that only ground shaking is primary, while others may include buildings falling down and soil liquefaction.) The answer also shows how one effect may lead to another, in that broken gas mains, fallen pylons and disrupted sewage systems may lead to fires and disease, respectively. Examples are well worked into the answer, particularly the reference to poorly constructed buildings in Gujurat. The reference to the time frame is very good, indicating that some secondary effects can last a long time, showing that the impact of an earthquake lasts much longer than the few seconds over which it occurs. All this material clearly fits the descriptor for Level 2 in the mark scheme. This answer would be awarded credit towards the top of Level 2 and would receive 7 marks.

(c) Discuss the effectiveness of the methods used to lessen the impact of earthquakes on the population of an area. **(10 marks)**

e The command 'discuss' requires you to present a verbal debate – some evaluation of effectiveness is needed.

Mark scheme

Level 1: simple statements of methods, in effect little more than a list of the ways people can attempt to lessen the impact. (1–4 marks)

Level 2: recognises that there can be categories, or shows the general aim behind the methods. Begins to show how these methods work and how effective they have been, perhaps with some small references to located examples. (5–8 marks)

Level 3: shows a clear indication that methods can be divided into categories and discusses the purposes of such attempts. Clearly makes critical evaluations of the methods with details on how they can be made to work. Links methodology and its effectiveness with clear references to located examples. (9–10 marks)

Student answer

Trying to predict earthquakes is very difficult. There have been some attempts based upon monitoring groundwater levels, the release of radon gas and unusual animal behaviour but these are very unreliable. Studies of fault lines and their previous seismic history sometimes reveal gaps where the next earthquake might strike. Investigations on the San Andreas fault in California revealed a gap at Loma Prieta and that is where the next earthquake occurred. Unfortunately, they could not predict when it would occur. People, though, could be more prepared as a result of such investigations. Prevention is probably impossible, although it has been suggested that oil could be poured down the San Andreas fault to lubricate it so that it moves consistently with no 'jerky' actions.

You can protect against an earthquake by building hazard-resistant structures (large concrete weights to swing to counteract the stress, buildings placed on rubber shock absorbers or adding cross-bracing to hold structures together when they shake). Older structures can be retrofitted with such devices. In the California earthquake of 1989, there were only 60+ deaths in an area that contained many earthquake-proof buildings, whereas in Armenia (1988) where buildings were often poorly built, a similar sized earthquake killed over 25,000 people, many as a result of the collapse of such buildings. 'Smart' meters can be installed which will cut off gas supplies in the event of a tremor at a certain scale. At a personal level, people can be educated to tell them what to do when the earthquake strikes and how to prepare supplies. The Japanese run earthquake drills in schools, offices and factories, and many countries issue lists of supplies that people should keep handy (torch, water, canned food, first-aid kit, can opener, matches, toilet paper). The authorities can help by keeping their equipment up-to-date, such as heavy lifting equipment, and having well-trained emergency services. In California, many local authorities carry out land zoning where they do not allow vulnerable buildings (schools, hospitals) to be constructed on the areas at greatest risk.

10/10 marks awarded. This is a comprehensive and excellent answer. Categories are recognised and detailed examples of methods are given within each category. The purpose of such methods is discussed with a critical evaluation of their effectiveness. This is backed up by clear exemplar material where similar sized earthquakes are compared. By looking at outcomes in California (1989) and Armenia (1988), similar sized events, the answer shows just how effective earthquake-proof buildings can be. All this material fits the descriptor for Level 3 in the mark scheme and would be awarded the maximum mark of 10.

Question 2 Essay question

The hazards presented by earthquakes and volcanic activity have the greatest impact on the poorest members of the world's population. To what extent do you agree with this view? (40 marks)

e Note the instruction 'To what extent'. You should give some indication of how far you think this statement is appropriate and valid. Note there is no correct answer — it depends on the evidence you present.

Appropriate content for a response to this question might include:

- the concept of a hazard
- an understanding of the two processes as hazards and the extent to which they impact upon human economies and societies
- the possibility of management
- areas at risk compared with the income/economic development of the people living there
- variations in the capacity to adapt to manage processes and impacts
- different impacts on different groups within the same population, such as the vulnerability of informal settlements
- case study material/exemplars

Synopticity could emerge with some of the following:

- a critical understanding of the processes that produce volcanic and earthquake hazard events and the context in which they are produced
- understanding the context of varying timescales (frequency etc.)
- an understanding of the impact of volcanic and earthquake events
- an understanding of the vulnerability of different populations to these hazards
- a critical understanding of the vulnerability of different regions, particularly an understanding of the differences between richer and poorer areas and the contrast between urban and rural environments
- understanding the capacity and willingness of people to deal with these hazards
- evidence of breadth/depth of case study material

The question requires a discussion and the response should come to a view. Any conclusion can be credited as long as it is measured and reasonable, and related to the content of the answer.

Student answer

Volcanic activity and earthquakes are both hazards and when they occur they put people's lives at risk. Not only this but they threaten their homes, livelihoods and businesses, apart from destroying part of a country's transport and energy networks. Fortunately, many people are not affected by such hazards as they do not live on or near a plate boundary where earthquakes and volcanoes generally occur. For those people that do, these hazards tend to occur infrequently. It would be unusual, for example, for a person to suffer two earthquakes in a lifetime. For those people who do live on the slopes of an active volcano, eruptions can be more frequent, but not always on the same side, for example Mount Etna.

Volcanic activity and earthquakes are the result of plate movement, either moving towards each other, away or sliding alongside. Volcanoes are formed when plates move apart or when one plate is subducted under another, such as in western

South America where the Nazca plate is going under the South American plate. Earthquakes can occur along any plate boundary and are the result of the plates sticking and then 'jerking' forwards. When such stress is released, parts of the surface suffer an intense shaking motion which lasts for just a few seconds.

It is impossible to stop such hazards, but people can prepare in order to lessen the impact upon themselves and the built environment. In earthquake-prone areas buildings can have weights added to them to counteract stress, be constructed with rubber shock absorbers and have some element of cross-bracing. Older buildings can be retrofitted with such devices. Education programmes can give people a better chance of survival by telling them what to do if an earthquake strikes and what things they should store in an earthquake survival pack (tinned food, water bottles, can opener, matches, torch etc.). 'Smart' meters have also been installed which will cut off gas supplies when affected by a certain strength of tremor. Authorities can maintain heavy lifting gear and have emergency services (fire, rescue, medical) trained and ready to act when an earthquake strikes. In California, the most hazardous areas in the event of an earthquake have been identified and certain types of buildings such as schools and hospitals are not allowed to be constructed there. Californian schools and hospitals have earthquake-proof features built into them. In areas of volcanic activity, volcanoes can be closely monitored and populations given adequate warnings. Authorities can also act to slow or divert lava flows by erecting barriers, dropping concrete blocks by helicopter and blasting openings in lava tubes. All of this was done in the eruptions on Mt Etna in the early 1990s.

All of the above measures cost a great deal of money, which makes the inhabitants of wealthier countries less vulnerable to such hazards. Volcanic and earthquake events, though, have a greater monetary cost to richer areas than to poorer ones, but in terms of the impact upon lives and an area's economy, you would have to argue the opposite. In the 1989 Loma Prieta earthquake in California, with its earthquake-proof buildings, there were only a small number of deaths, whereas in Armenia in 1988, an earthquake of similar size killed over 25,000 people, many of whom were crushed in buildings that had soft foundations and weak structures. Tens of thousands of people were killed in the Gujurat earthquake of 2001, many when poorly constructed buildings fell on them. In developing countries, there is also a difference in the way richer people can deal with such events as against people who live in shanty towns. The impact on poor people can last a long time as they struggle to rebuild their lives with very little help from the authorities. Help is often needed from richer countries in terms of medical supplies, lifting gear and personnel trained to help in such circumstances. They may also require a great deal of help in order to try to rebuild some of their economy.

It does not always work in such a way. Many buildings were destroyed and there was a large death toll in the Kobe (Japan) earthquake (1995) from falling buildings, fires and disease, which was surprising as Japan, being developed, was supposed to be in the forefront of earthquake management. The Mt Etna eruptions of the early 1990s might have been contained, but later eruptions destroyed many buildings in one of the ski stations on the volcano's side, although there was no loss of life.

In 2009, there was a major earthquake in L'Aquila in Italy, another developed country. Nearly 300 people were thought to have died, with many thousands being

made homeless. The earthquake devastated an area where there is wealth. In some ways the wealth helped in that the response was quick with rescuers from all over Europe; but many people are still without homes and many others are being forced to live away from the area in hotels on the coast. A rich country may be able to cope with such a disaster, but it is the individuals who have to suffer the most, irrespective of their own wealth.

In most circumstances, though, it is true to say that through having money, either personally or through a richer state, it is possible to lessen one's vulnerability to volcanic and earthquake hazards. People in developed areas can also take out insurance to cover their losses, although this can be very expensive for individuals. In a Californian earthquake, federal aid and other forms of help are almost always available. Aid can be supplied to poorer countries, particularly in the short term, but longer term help is much less readily available.

35–36/40 marks awarded. This is an excellent answer. There is strong evidence of detailed and accurate knowledge of the two hazards and the extent to which they impact upon human societies and economies. Clear understanding is shown of how such hazards can be managed but also of how people are vulnerable to such events. The answer also shows a good critical understanding of vulnerability based upon people's wealth. Good use is made of exemplar material both to agree with the view expressed and also to question whether it is always true. Case studies therefore show evidence of both breadth and depth, although there is still room for further supportive material.

There is a high level of insight and the writer shows an ability to identify and interpret a wide range of material. Much of the material is synoptic and there is a clear element of discussion present. The answer comes to a distinct conclusion and although there is wide agreement with the view put forward, the writer does question, with exemplar material, whether this is always the case. The commentary on the Italian example illustrates a degree of flair. All of this fits very well into the descriptor for Level 4 in the mark scheme. Such an answer, with detailed and accurate references to case studies and being fully synoptic, would be awarded a mark towards the middle of the level (35 or 36 marks).

Weather and climate and associated hazards

Question 3

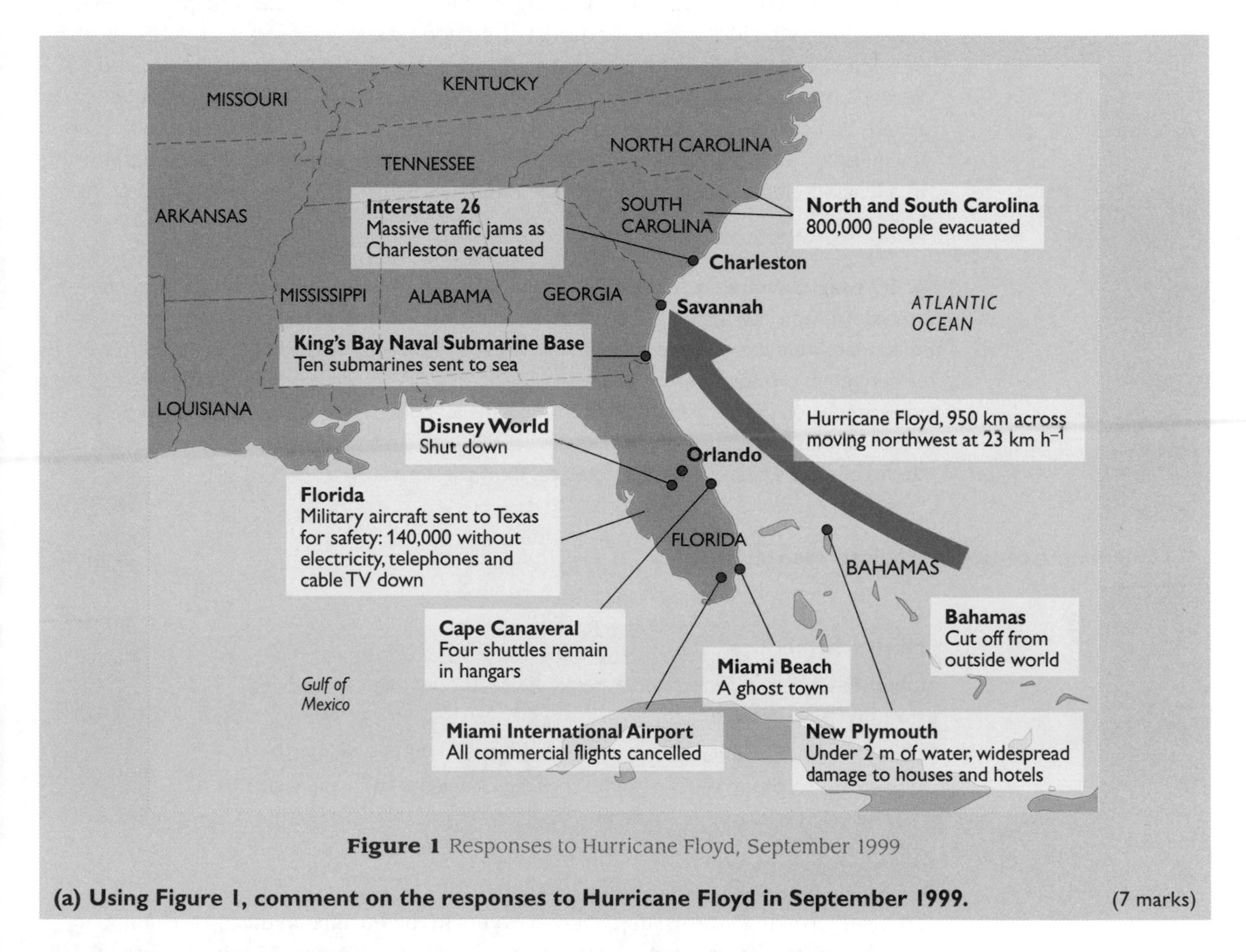

Figure 1 Responses to Hurricane Floyd, September 1999

(a) Using Figure 1, comment on the responses to Hurricane Floyd in September 1999. (7 marks)

Note the command 'comment on'. You should make an observation that is relevant, geographical and appropriate to the responses that is not immediately obvious.

Mark scheme

Level 1: basic/simple statements describing the impacts of the storm with passing reference to responses. Comment may be limited to the acknowledgement that the USA is a more developed country. (1–4 marks)

Level 2: the answer focuses on the responses to Hurricane Floyd and there is clear comment. (5–7 marks)

Student answer

Hurricane Floyd swept along the east coast of the USA, affecting the states of Florida, Georgia and North and South Carolina. There was obviously enough warning as people from coastal settlements, such as Miami Beach and Charleston were evacuated before the storm hit. The military forces were able to protect their planes and submarines by sending them away, outside of the path of the hurricane. Planes were evacuated to Texas, and submarines sent out into the Atlantic. Commercial flights were also cancelled at Miami airport, to protect both planes and passengers. Disney World in Florida, a major tourist attraction was also closed. As the USA is a developed country it has the weather forecasting technology to predict when and where a hurricane will hit so that people and businesses can be warned in advance. Also it has evacuation procedures in place which can be quickly put into practice, so few people die. However the economic cost of this event will be much greater than in a less developed country such as the Bahamas, which was also in the path of the hurricane.

e **6/7 marks awarded.** The candidate provides a rather simple description of the responses to the storm with information lifted straight from the map. It would have been improved by using the evidence from the figure to suggest how severe this particular tropical revolving storm was. The comments provided only include reference to the USA's more economically developed status and expand upon this in relation to the ability of the authorities to implement strategies designed for such an event. This answer would access Level 2, with 6 marks — more commentary, possibly in terms of wider impact, is needed to access the highest mark.

(b) Explain the causes of tropical revolving storms. (8 marks)

Mark scheme

Level 1: simple and generalised statements of causes of tropical revolving storms with no depth or detail. (1–4 marks)

Level 2: specific and detailed causes of tropical revolving storms. Better answers may relate to specific events, showing knowledge of locations such as the Gulf of Mexico. (5–8 marks)

Student answer

Tropical revolving storms are common events in the Caribbean during the summer months, when it is hot, and the sun is directly overhead. The ITCZ moves north at this time of the year and drags with it a belt of low pressure. In low-pressure weather systems hot air rises, and then cools and condensation occurs, causing clouds to build up. When convection heating occurs over warm tropical water rapid uplift of warm moist air causes a tropical storm to build. A distance away from the equator, (from about 5 degrees upwards) the Coriolis force causes the rising air to spin. Surface winds are the result of air being drawn towards the centre of the low to replace the air that has risen.

5/8 marks awarded. This response is focused well enough on the causes of tropical revolving storms but it lacks the detail necessary for a top mark as it fails to relate to specific events or locations. The language used is a little simplistic but the facts presented are accurate so it would achieve a low Level 2 score of 5 marks.

(c) With reference to one tropical region that you have studied, describe and explain the characteristic features of the climate of that region. (10 marks)

Note the two commands here: describe and explain.

Mark scheme

Level 1: simple description of the climate of one type of tropical region. (1–4 marks)

Level 2: specific detail is used to describe the regime of one tropical climate, with an attempt to explain one aspect (temperature/precipitation/winds). (5–8 marks)

Level 3: fully developed answer with both accurate description and clear explanation of the climate of one appropriate tropical region. (9–10 marks)

Student answer

The equatorial climate is hot and wet all year long and has no real seasonal pattern in its temperatures and rainfall. Average monthly temperatures are between 26 to 28 degrees Celsius, and they vary more between day and night than they do monthly. Rainfall is usually over 2,000 mm p.a. more or less evenly spread throughout the year, although there might be a short dry season as you move further from the equator. Air pressure is generally low, leading to daily convectional storms in the late afternoon, often accompanied by thunder and lightning. It is not a very windy climate and the length of the day and night is roughly the same all year through.

These conditions occur because air pressure is low between 5 degrees N and S of the equator, as along the ITCZ warm air rises due to the intense overhead sun. As the air rises it cools and water vapour condenses, forming cumulonimbus clouds and heavy rain. It is hot throughout the year because the sun is always more or less overhead.

6/10 marks awarded. The candidate selects the equatorial climate region, which is an appropriate choice. Description of the climate is sound, covering temperature and precipitation patterns with knowledge of values for each of these elements. Explanation of the climate is somewhat limited, referring accurately to latitude but demonstrating a partial understanding of the role of the ITCZ in relation to rainfall and pressure. There is no attempt to link the answer to specific localities (other than 5°N and S of the equator) so this answer would achieve Level 2: 6 marks. To improve, greater sophistication of understanding of the processes involved would be needed.

Question 4 **Essay question**

'Urban areas have a significant impact on their climatic characteristics.' Discuss this statement. (40 marks)

ⓔ Make sure you read every word of the question — the term 'significant' is crucial here. It is not just a question on urban climates.

Appropriate content for a response to this question will include:

- reasons for the existence of the urban heat island/heat dome
- reference to humidity, cloud cover, and thunderstorms and precipitation within cities, and reasons why variations occur between urban and rural areas
- differences in wind speed between rural and urban areas, turbulence, funnel/Venturi effect within cities
- air quality, fog, photochemical smog, causes of pollution

Synopticity is achieved when there is a critical appreciation of the varying effect of cities on local climates around the world by:

- location in relation to latitude, relief, distance from the sea and general climate experienced
- population size and extent of the built-up area
- level of industrial development and type of economic activity present, particularly when comparing countries in the developing world with those in the developed world
- environmental legislation, again comparing developing nations with those in developed countries

The question requires some discussion and candidates are expected to produce a reasoned, measured conclusion, which relates clearly to the preceding content.

Student answer

It has been widely recognised for a number of decades now that the climate of large urban areas varies significantly from the surrounding rural areas. The urban heat island is the most well-known aspect of this, with temperatures in many city centres much higher than the surrounding countryside areas, but other aspects of climate, such as wind, rainfall, cloud cover and air quality also show variations too.

Generally speaking, the larger the size of the city the more impact it will have on the climate. However, variations to this rule do apply as big cities in tropical latitudes, where it is already very hot, may not show as much variation as those in mid-latitudes, so somewhere like New York or Paris might have more difference than somewhere like Singapore, closer to the equator.

In developed countries, where there are lots of cars, factories and high-rise buildings that are heated in winter and air-conditioned in summer, there might be higher city temperatures but poorer air quality than in developing countries. However, in some cities in the developing world, such as in Beijing, the air quality is poor because lots of coal is burned. This gives rise to smog when the particulates act as condensation nuclei. Back in the 1950s London used to have a lot of smog but clean air legislation stopped people burning coal in their homes, so this does not occur these days.

As a rule, dark urban land surfaces, such as tarmac and concrete and glass absorb radiation during the day and give this out at night. The difference in the temperature between city and countryside is most noticeable during the night, particularly during the winter months. Because of this cities are far less likely to have frosty mornings than rural areas.

In some cities, where the streets are in a grid pattern and the buildings are high rise, wind speeds will be faster as the air is channelled and funnelled in the direction of the roads. Chicago is known as the windy city because of this. However, the buildings can also act as a windbreak and so it can be calmer in the city during a storm than in the surrounding countryside.

Industrial cities have been known to be wetter places in some countries, usually when the air pressure is low and the air is rising. Manchester and Preston in the northwest are well known for this. Preston rain soaks you right through to the skin and goes on for hours once it starts, believe me. Factory chimneys belch smoke into the air, smoke particles act as condensation nuclei and raindrops form around these. Even just downwind of the city rainfall levels have been shown to be higher.

Poor air quality can lead to photochemical smog in some cities around the world. This type of smog creates a yellow haze over the city which is full of ozone and can be bad for people's health. Car emissions and fumes from power stations burning fossil fuels react with sunlight to produce this smog. This is found mostly in sunny places like Los Angeles, Mexico City and Athens, where it is very hot and calm during the summer and there is lots of traffic. Because the air is still there is little chance of the smog being blown away so it builds up and hangs over these cities for weeks at a time. We don't get much of it in England, although the air quality has been shown to be as bad as smoking 60 cigs a day in Oxford sometimes. In many cities like Athens new laws have been brought in to deal with this. People with odd number plates can only drive their cars into the city every other day and even numbers on the remaining days. However in many developing cities like Mumbai air quality is actually getting worse as they continue their industrial development and cities sprawl due to shanty towns growing on the urban fringe.

I think that I have shown that overall cities have a significant impact on climatic characteristics.

e 16/40 marks awarded. The answer is relevant and accurate, and shows some understanding, even though some of the ideas are simply stated (without much maturity of style). Unfortunately it is too brief. A range of ideas is expressed, demonstrating some understanding of concepts, and the examples are varied, despite being brief and not fully developed. The student knows something at least about the theme of the question, and tackles a range of aspects of the impact of cities on climate. However, the importance of the extent of impact has not really been addressed. Overall the breadth of ideas expressed balances out the immaturity of expression, so a middle-range mark within Level 2 is awarded.

Ecosystems: change and challenge

Question 5

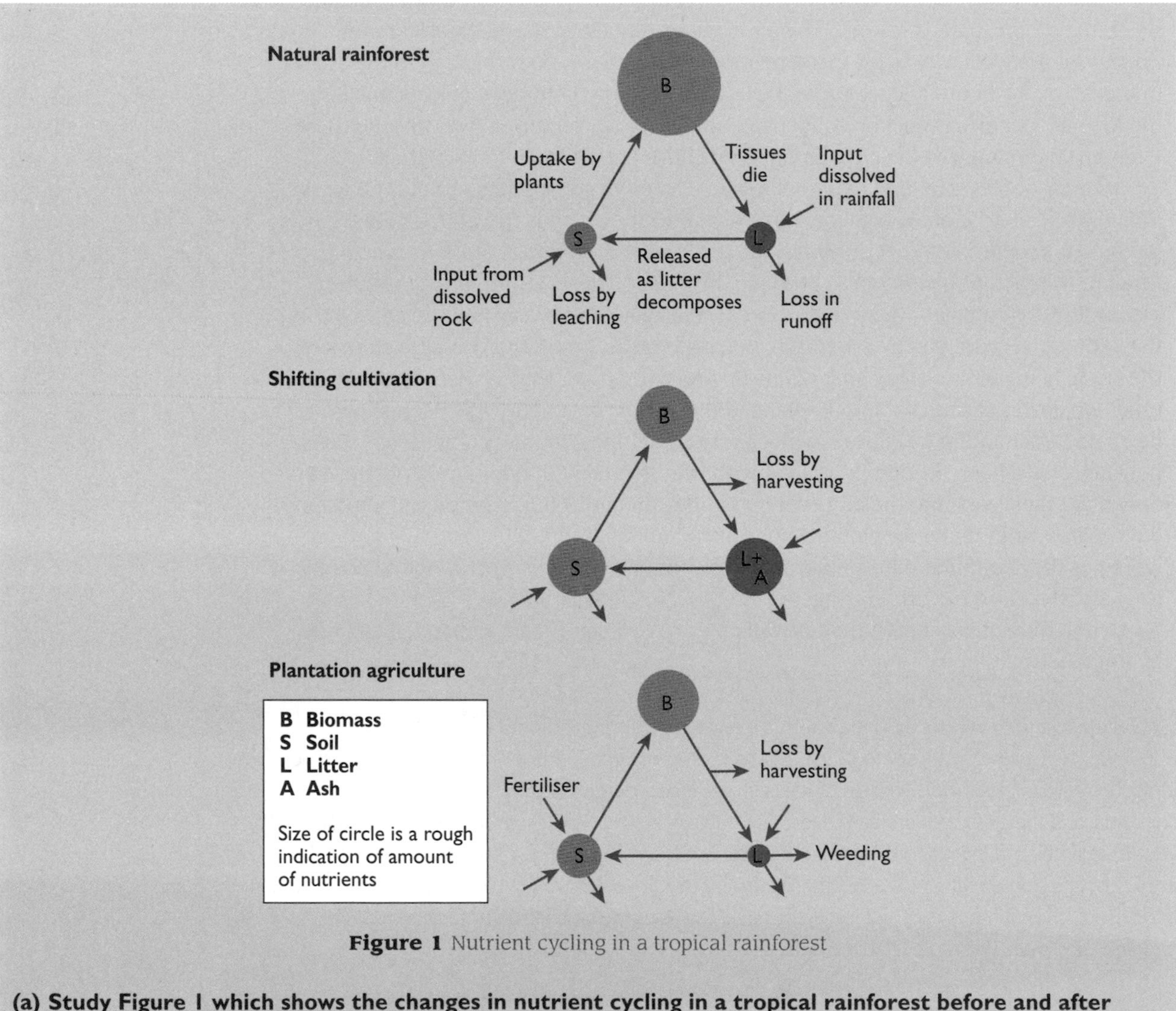

Figure 1 Nutrient cycling in a tropical rainforest

(a) Study Figure 1 which shows the changes in nutrient cycling in a tropical rainforest before and after human activity has taken place. Describe the changes that have taken place. (7 marks)

Make sure you refer explicitly and precisely to the stimulus material.

Mark scheme

Level 1: basic/simple but incomplete statements stating the differences between the undisturbed forest and at least one of the other nutrient cycle diagrams. (1–4 marks)

Level 2: clear description of the differences between the nutrient stores in the undisturbed TRF and both shifting cultivation and plantation agriculture. (5–7 marks)

Student answer

In the natural rainforest most of the nutrients are stored in the biomass and there are only small amounts in the litter and the soil because nutrients are transferred very quickly. Litter is produced from leaf fall all year long but is quickly broken down into humus and then taken up rapidly by the plants and trees from the soil. There are hardly any nutrients lost from the soil by run-off and leaching.

When the land is cleared for shifting cultivation the result is less storage of nutrients in the biomass, but a greater proportion in the soil and litter (due to the addition of ash) than before. Overall, there are fewer nutrients in total because after the trees are cut down and burned some nutrients are washed away into rivers and some are leached through the soil.

Plantation agriculture results in a smaller store of nutrients in the biomass and less nutrients than before too in the litter. Although some nutrients have been added to the soil by fertilisers, even more are lost due to harvesting and weeding.

e **7/7 marks awarded.** This is a full and clear response, which describes the differences between the nutrient stores in the rainforest and both shifting cultivation and plantation agriculture, so all 7 marks available would be awarded. The student clearly understands both the diagrams and the concepts involved.

(b) Choose one biome of one tropical region that you have studied.

(i) Describe the main characteristics of that biome. (8 marks)

Mark scheme

Level 1: a basic description of one or more elements of the chosen biome/ecosystem. (1–4 marks)

Level 2: a clear description, using some accurate facts, relating to vegetation, climate and soils. (5–8 marks)

Student answer

The tropical rainforest biome has an equatorial climate, which is hot and wet throughout the year. Typically the climax vegetation consists of layers of trees; the tallest called emergents can reach up to 45 metres high and poke out above the main canopy layer which is about 5–10 metres lower. The yearlong growing season produces the densest forest found on Earth, with more species of plants, animals, birds etc. than anywhere else in the world. Trees are mainly deciduous, e.g. teak and mahogany, and from the 400+ species that grow some are always in leaf, flower and fruit, so there is plenty of food for wildlife. It is very dark in the lower layers of the forest so there is little undergrowth, apart from in clearings or along river banks. Fungi quickly break down dead and decaying plants and trees.

In this biome the soil is mainly red latosol and this is very deep, but is not fertile as the plants quickly use up any nutrients that are input from the litter. It is so deep because the hot and wet climate leads to rapid chemical weathering of the underlying rock.

e 6/8 marks awarded. The choice of the tropical rainforest biome is appropriate and the candidate provides a sound description of the vegetation, with some specific detail. However, the student has not recognised that the question requires a description of the biome — there is more to write about. For example, description of the soil and climate is more basic, so the answer would reach mid-Level 2, with 6 marks.

(ii) Discuss the ecological responses to the climate by both plants and animals. (10 marks)

e Note the command 'discuss'. You are expected to demonstrate an element of debate in your answer.

Mark scheme

Level 1: simple adaptations, that describe either the ways that the vegetation or the animals have evolved to survive in the climate of the selected biome. (1–4 marks)

Level 2: a clear response which describes adaptations of both vegetation and wildlife to the climate of the chosen biome. (5–8 marks)

Level 3: at this level expect to see detail in the response, particular species of vegetation or animal might be used accurately to discuss the ecological responses to the climate. (9–10 marks)

Student answer

The plants and animals have adapted to the climate in many ways. Some plants, called epiphytes, actually grow on the branches initially, and wrap themselves around the tree for support as there is no sunlight on the forest floor. The trees have few branches on their trunks because they shoot up quickly towards the light when a gap in the forest appears, only branching out when they reach the light. Some trees have buttress roots to give them support at the base as they are so tall. Many trees have developed leaves that can cope with the heavy rainfall, by developing drip-tips so that water does not collect on them. Tree trunks are smooth and the bark is thin as there is no need to protect them from frost. The tallest trees, called emergents have flexible branches (e.g. balsa) which do not break in strong winds that can occur above the forest.

Most of the animals live in the trees, where food and light are available. Snakes wrap themselves around the branches; jaguars and other predators lie in wait for their food, taken from the thousands of bird and smaller animal species, such as parrots and monkeys. Some animals have become specially adapted to live in trees, sloths hang from the trees and some monkeys have extra long tails to help them swing between branches. Because the forest provides an abundant supply of food all year round it supports the greatest variety of animal and plant life on earth.

6/10 marks awarded. The command word 'discuss' requires the candidate to make good use of evidence and appropriate examples, and to present an argument — perhaps, for example, that the vegetation is influenced by the climate in the first place but the subsequent trophic levels, in particular herbivores, are primarily influenced by the food available to them. This answer also lacks depth of understanding and reference to locations, such as the Amazon basin. It does demonstrate an understanding of some of the ways in which the vegetation has adapted to the hot wet climate, although the animal life is covered less convincingly, and it lacks the use of specific species in support. Overall, the answer would access 6 marks and is judged to lie in the middle of Level 2.

Question 6 **Essay question**

'Development, biodiversity and sustainability are incompatible goals.' Discuss this statement in the context of one tropical biome. (40 marks)

There are three key words in this question — development, biodiversity and sustainability. Make sure you use these in your answer regularly. It tells the examiner that you have recognised the focus of the question.

Appropriate content for a response to this question will include:

- an understanding of the original biodiversity of the tropical biome selected, and an appreciation of the levels of biodiversity of the main species of vegetation, animal, insect, birds etc. at the present time in the specific biome (the tropical rainforest, the savanna grasslands or the tropical monsoon or mangrove forest)
- an understanding of the concept of sustainable development, meeting the needs of the present without damaging the ability of future generations to meet their needs, so allowing development to take place but protecting the environment at the same time
- reference to supporting case study material/locations where development is occurring. Such examples may be used in support of the statement, or may be used to challenge the statement that development, biodiversity and sustainability are actually compatible goals

Synopticity will be achieved when there is:

- a consideration of the roles and attitudes of different groups in relation to development, perhaps between indigenous people, environmentalists and those who wish to develop the biome for economic gain

- evidence in the depth of the chosen case-study material
- detailed critical understanding of the concept of sustainable development
- recognition that attitudes towards sustainable development change over space and time
- understanding of the role of both developed and developing countries in solving the global nature of the constraints on sustainable development, by indirect human activities, such as climate change

This question requires discussion and a reasoned point of view should be established in the conclusion. The conclusion itself must relate to the preceding content.

Student answer

The tropical rainforest is the most productive and diverse biome on Earth. In the Amazon rainforest alone there are more than 50,000 known species of plant, with hundreds of species of mammals, reptiles and fish and thousands of species of birds and insects. This biome is found between 10 degrees N and S of the equator where the climate is constantly hot and wet, but it is thought that only 50% of the original rainforest now remains, due to deforestation. Very little original forest remains in Africa and Asia and if present rates of clearance continue there will be hardly any primary forest remaining in 100 years time.

Removal of tropical rainforest destroys habitats for wildlife and disturbs the food chain, which has a knock-on effect on primary producers through to herbivores and carnivores. It has been estimated that thousands of species, including primates such as the orang-utan, are threatened. It is also important that rainforests are protected because when plants photosynthesise they produce oxygen, needed for the existence of life on Earth. Also trees act as a carbon sink and help to absorb carbon emissions produced from the burning and combustion of fossil fuels. Global warming is forecasted to be a major concern for this century and continuing to cut down rainforests will only contribute further to this.

Development of the tropical rainforests worldwide occurs for many reasons. In the far east, in countries such as Malaysia and Indonesia, economic development continues at a rapid pace and wood is needed for building, to sell as an export, and the land is also required for settlement and industrial development. In such places population pressure also requires land to be cleared for agricultural and HEP production. In Indonesia and Brazil re-settlement programmes have encouraged people and offered financial incentives to move from densely crowded cities, such as Jakarta and São Paulo to newly cleared areas in the rainforests.

In some parts of the world little consultation is made with locals when the land is cleared. For example in the Congo and Nigeria, where oil has been discovered, TNCs have forced locals from their villages and have destroyed the land. In Brazil vast areas have been cleared for mining of iron ore around Carajas, and huge cattle ranches and plantations growing cash crops such as palm oil and sugar cane have been developed in regions such as Rondonia. It is highly unlikely that these areas will ever regenerate to the highly diverse ecosystems they once were.

Sustainable development allows present generations of people to use the rainforests so long as the environment is protected for people in the future, so that they can continue to enjoy and benefit from the land. There are ways that the rainforest could be managed more sustainably and global agreements have been made to try and ensure that this happens.

The 1997 Kyoto Treaty included an agreement to protect global diversity in addition to the well-known plan to cut CO_2 emissions. Also in 2002 key issues relating to species diversity were discussed at a summit in Johannesburg and the goal agreed was to halt species extinction. Although there is global intent and goodwill, only a few countries, mainly the wealthier ones, such as the UK, have actually started to implement changes agreed. In many developing countries it is almost impossible to ensure that environmental agreements are implemented on a national scale as they do not have the infrastructure or money to do so.

On a smaller scale, however, successful efforts at managing diversity can be seen. Ecotourism, selective logging and traditional methods of land management such as rubber tapping and shifting cultivation do not damage the environment and allow for regeneration. In some areas of the rainforests, protected areas and National Parks have been created. For example, the Central Amazon Complex in Brazil, more than 6 million hectares in size, contains one of the most diverse wildlife communities in the world. A very small indigenous population of Amerindians live a traditional way of life in this remote part of the Amazon rainforest. The land within the complex is split into four zones for management purposes, from the primitive zone where the land is untouched and has the most protection through to the intensive use zone where some economic activity, such as rubber tapping, is permitted. Local communities have worked together with the authorities and other development agencies to agree on a management strategy which will ensure continued biodiversity.

To conclude, large-scale development of the rainforest for short-term economic gain can never be compatible with biodiversity and sustainability. So long as the will to protect is there, small-scale developments can however be managed appropriately to allow for present and future generations to benefit from their environment and to allow biodiversity to flourish.

e **30/40 marks awarded.** This answer would access Level 3 and 30 marks. It demonstrates frequent evidence of detailed and thorough understanding, and it is reasonably sophisticated and mature in style. The argument is direct and logical, and there is strong evidence of synthesis. The essay is purposeful — an argument is assembled, and the student ends with a conclusion based on the argument established. There is evidence of critical understanding, for example the paragraph on the Kyoto Treaty.

So, why was Level 4 not awarded? Because examiners are looking for greater insight and depth, possibly with a degree of 'flair'. The last section on ecotourism is not detailed enough. This part of the answer is where the student is being positive about biodiversity and development existing hand in hand, referring to the importance of small-scale developments that are sustainable. There could have been more detail here, and a degree of flair of knowledge or commentary (even opinion) about a small-scale enterprise he/she had studied.

World cities

Question 7

Study Figure 1 which shows the global distribution of two types of city.

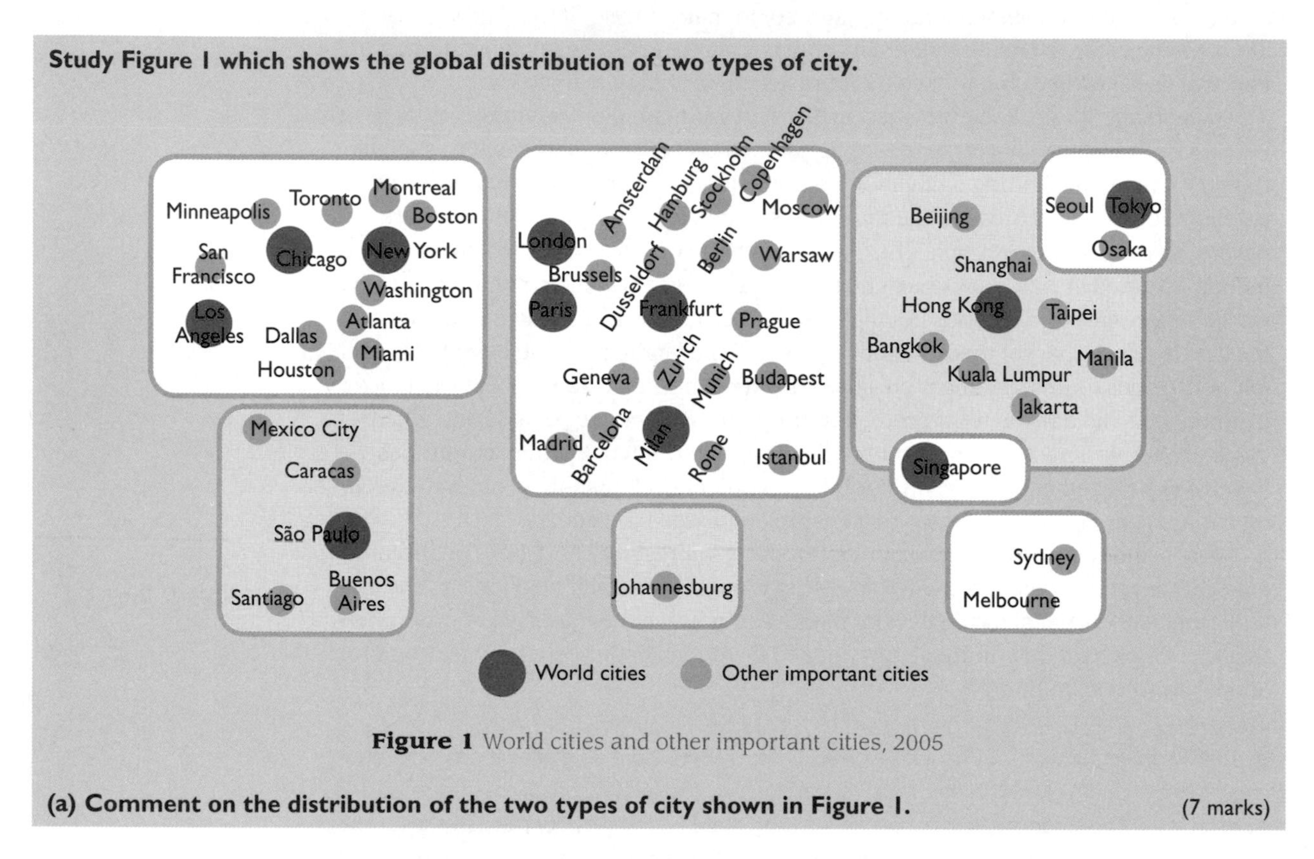

Figure 1 World cities and other important cities, 2005

(a) Comment on the distribution of the two types of city shown in Figure 1. (7 marks)

ⓔ Note the command word 'comment on', and the key word 'distribution'.

Mark scheme

Level 1: simple listing of cities by continent; recognition that there are variations in the totals of cities by continent. Basic recognition of types of city shown and their distribution. Commentary is lacking or simplistic. (1–4 marks)

Level 2: commentary that reflects on the distributions shown that may suggest some contributory factor, e.g. level of development. Critical comments on the data shown. Overall a more sophisticated response. (5–7 marks)

Student answer

There is a variation in the distribution of world cities in the world as shown on Figure 1. There are four in Europe, three in each of North America and Asia, and one in South America. There are none in Africa or Australasia. There are also more other important cities in Europe and North America. There is only one large city in Africa, Johannesburg. Why is Cairo not included here? In fact the more I look at these data the more I think they are not very helpful. For example, Manchester does not feature and it is as large as Prague, and there are many more large cities in China. Overall I think these data are simplistic as they are trying to make a comment that there are more large and important cities in more developed countries, but this is not the case as there are far more people in countries that are less developed. In fact no Indian city is mentioned, which is wrong.

e **6/7 marks awarded.** The command 'comment on' is widely used in A2 examinations, especially where stimulus material is involved. Here, the data are relatively simple and initially appear to support the view that there are more 'world cities' in the developed world than in the developing world. This student begins to describe the data in very simplistic terms (and gains a series of Level 1 marks), but it is only when he/she recognises that Cairo is missing that the data begin to be questioned. The command word 'comment on' invites the student to make a statement that can be inferred from the data. This student concentrates on being critical of the data (demonstrating critical understanding), and hence accesses Level 2: 6 marks.

(b) With reference to one or more examples, describe how urban growth can cause social and economic problems in urban areas. (8 marks)

e Note this question can be answered with either 'depth' — one example; or 'breadth' — more than one example.

Mark scheme

Level 1: generalised account of problems that could refer to the growth of any city/urban area in the world. Problems tend to be listed simplistically rather than dealt with in depth; or discussion of one, or one type of, problem only. (1–4 marks)

Level 2: more than one problem discussed. Specific statements relating to a named city/urban area access this level. Answers are detailed, have depth and are more sophisticated. Higher mark responses should refer to both social and economic problems. (5–8 marks)

Student answer

Urban growth is taking place in a variety of forms around the world. In developing countries such as Egypt, there is huge pressure on the capital Cairo. In developed countries such as the UK, where counter-urbanisation is taking place, there are also problems being created.

Cairo is built on raised land (the Magattam Hills) close to the River Nile, and is located between the fertile farmlands of the Nile Delta to the north and the irrigated land alongside the Nile to the south. It has very high population densities (over 32,000 per km^2) and a high infant mortality rate of 105 per 1,000 live births. The rapid growth of the city has caused some problems. Thirty per cent of the city has no public sewerage system. Fifty-five per cent of waste water is untreated as it travels through open canals and rivers to the sea. Although the city does not have many extensive areas of squatter settlements (examples include Bulaq and Chobra), many people live in inappropriate locations — the Cities of the Dead (the tombs of old Cairo) and on rooftops in makeshift dwellings. There is serious air pollution, caused by traffic and open-air cooking stoves. Waste disposal is disorganised — in some areas it is done by the zabbaleen with donkey carts. Most of the above are social problems, but will have economic consequences.

On the other hand in the UK, there are many areas in the southwest where people are moving to retire, into small villages and towns such as Ottery St Mary. By definition, such people are older and will get even older. This will cause social problems in terms of having an imbalance in the population towards older people and so there will need to be greater provision for these older people — more retirement homes and more sheltered accommodation. There are also more housing developments for people who are not ready yet to go into sheltered housing. This puts pressure on local services such as transport and health. As with Cairo above, the consequences of these will be economic as local authorities have to plan to meet these requirements, both short and long term, and at a cost.

e **7/8 marks awarded.** The student has satisfied all of the requirements for Level 2 — both types of problems have been discussed, and in the context of two different areas and in terms of the impact of two different urban processes. Some parts of the answer are specific, especially that related to Cairo, whereas other aspects progress little beyond the general. More precise problems facing Ottery St Mary would have raised the quality of the answer. The answer could also have been improved if the student had been more explicit about the nature of 'social' problems and 'economic' problems. The examiner is having to do much of the work on behalf of the student. Despite this, Level 2 credit is awarded: 7 marks.

(c) Discuss the attempts by planners to reduce the impact of cities on the physical environment. (10 marks)

e The command 'discuss' requires you to present a verbal debate — some evaluation is needed.

Mark scheme

Level 1: limited knowledge and understanding of how planners have attempted to reduce the impact of cities on the physical environment. Cause and effect are not well understood and there is limited use of examples. Limited analysis of the attempts by planners to reduce the impact of cities on the physical environment. (1–4 marks)

Level 2: some knowledge and understanding of how planners have attempted to reduce the impact of cities on the physical environment. Cause and effect are understood and some examples are given. Some analysis of the attempts by planners to reduce the impact of cities on the physical environment. Limited (if any) evaluation. (5–8 marks)

Level 3: detailed knowledge and understanding of how planners have attempted to reduce the impact of cities on the physical environment. Cause and effect are well understood, and there is effective use of detailed examples. Clear analysis and effective evaluation of the attempts by planners to reduce the impact of cities on the physical environment. (9–10 marks)

Student answer

In Cairo, planners have tried to reduce the impact of the city on the physical environment by dealing with some of the issues that the growth of the city has caused for the environment. These include dealing with sewage waste, traffic congestion and air pollution and spontaneous settlements which use up large areas of land.

There is a project to repair the existing sewers and to extend the sewerage system to those parts of the city currently not served. This includes the provision of pipes carrying clean water into the Cities of the Dead and to the rooftop homes, and therefore allowing waste to be removed by the movement of water. There are proposed schemes to reduce the numbers of vehicles on the roads, though none have been successful. There are longer-term plans to extend the underground metro system to reduce the amount of traffic, thereby reducing the air pollution of the city. To reduce the amount of land being given over to spontaneous settlements, there has been the provision of low-cost accommodation in high-rise flats in new satellite towns in the outlying desert such as '10th Ramadan' and '15 May'. All of these should have beneficial effects on the physical environment.

Elsewhere in the world, Curitiba has a planned transportation system, which includes lanes on major streets devoted to a bus rapid transit system. There is only one price no matter how far you travel and you pay at the bus stops. This inexpensive, speedy transit service is used by more than 2 million people a day. There are more car owners per capita than anywhere in Brazil, and the population has doubled since 1974, yet traffic has declined by 30%, and atmospheric pollution is the lowest in Brazil.

The city has also paid careful attention to preserving and caring for its green areas, boasting 54 square metres of green space per inhabitant. Curitiba is now referred to as the ecological capital of Brazil, with a network of 28 parks and wooded areas. Residents have planted 1.5 million trees along city streets. Builders get tax breaks if their projects include green space. Flood waters diverted into new lakes in parks solved the problem of dangerous flooding, while also protecting valley floors and riverbanks, acting as a barrier to illegal occupation, and providing aesthetic and recreational value to the thousands of people who use city parks.

There is a 'green exchange' employment programme where low-income families living in shanty towns unreachable by truck bring their rubbish bags to neighbourhood centres, where they exchange them for bus tickets and food. This means less city litter and less disease, less rubbish dumped in sensitive areas such as rivers and a better life for the undernourished poor. There's also a programme for children where they can exchange recyclable waste for school supplies and chocolate. Under the 'garbage that's not garbage' programme, 70% of the city's trash is recycled by its residents. Once a week a lorry collects paper, cardboard, metal, plastic and glass that has been sorted in the city's homes. The city's paper recycling alone saves the equivalent of 1,200 trees a day. As well as the environmental benefits, money raised from selling materials goes into social programmes, and the city employs the homeless and recovering alcoholics in its garbage separation plant.

10/10 marks awarded. Planning attempts aimed at reducing the impact of cities on the physical environment include:

- controls on air pollution, especially from motor vehicles
- recycling waste and reductions in solid waste going to landfill
- reclamation of derelict land
- tackling traffic congestion, and
- the development of sustainable cities

As stated in the mark scheme, the key here is to provide both depth and detail of a range of management strategies, with good use being made of case studies. To gain the highest level, some statement of success or otherwise should be given.

This is an excellent response, making very good use of case study material in two separate locations. The latter, based on Curitiba, clearly takes the response into the concept of sustainable cities. Note that the student chooses not to make a concluding statement regarding the success of these schemes, but there is effective evaluation of several of them throughout the response. This is an acceptable way to approach this type of question. It is difficult to fault this answer — Level 3, 10 marks.

Question 8 **Essay question**

With reference to examples, discuss the overall effectiveness of urban regeneration schemes. (40 marks)

The command 'discuss' requires you to present a verbal debate — some evaluation is needed.

Appropriate content for a response to this question might include:

- a description of the issues facing identified areas prior to regeneration
- a definition of the term 'urban regeneration'
- an understanding of the purposes of urban regeneration

- a discussion of at least one urban regeneration scheme and of how it impacts on the identified areas affected
- a comparison of the different strategies adopted by identified areas

Synopticity is therefore achieved by:

- evidence in the breadth/depth of case study material
- detailed critical understanding of the issues facing the areas identified
- detailed critical understanding of the regeneration strategies in the areas identified
- a recognition of the importance of values and attitudes, and of the role of decision makers
- evaluative comments as to whether the schemes are/were successful

The question requires a discursive approach and the response should come to an overall view. Any conclusion can be credited as long as it is reasonable and related to the preceding content and argument.

Student answer

Urban regeneration schemes have been a common feature of most towns and cities in the UK in the last 20 years. Various governments have tried to regenerate cities over this time in a variety of ways. I am going to write about three such schemes, in London, Bradford and Southampton over this time period and discuss their overall effectiveness.

UDCs (Urban Development Corporations) are a form of property-led regeneration established in the 1980s under Margaret Thatcher, which were run by an executive board and were given money by central government to spend in the best way for the local area, although their aim was always to improve the area in such a way that businesses would see it as a good business opportunity. They were market led and property led because they made physical changes, for example they improved infrastructure to attract businesses (property led) and it was market forces, not planners, that decided the ultimate layout of the area — they wanted businesses to lead the way (market led).

An example of an Urban Development Corporation (UDC) is the London Docklands UDC (includes such as Wapping, Tower Hamlets and Lime House). They aimed to improve the area in such a way so as to attract any business (such as financial and media based) and they did this by improving the infrastructure (e.g. with the river bus and 90 km of new roads) and the environmental image of the area. In a way the UDC was very effective. It had a leverage ratio (in billions of pounds 1 to 8.5) and it created a second CBD on the Isle of Dogs (100% of which is let) which many newspapers, e.g. the *Telegraph*, located on. The area looked a lot better with new waterfront walks and parks and an ecology site at East India Dock basin.

However, the social aspect of regeneration wasn't looked at and the lives of local people were not improved e.g. the new river bus and Jubilee line were too expensive, and the jobs created were either relocated from elsewhere or were management jobs, unsuited to unemployed manual labourers from the former docks (only 17,000 were totally new).

With a population of over 450,000, Bradford (Yorkshire) is one of the ten largest cities in Britain. Its early growth was as a centre of the wool and textile trade. The city is famous for its links with the Bronte sisters and the artist David Hockney. It is also renowned for the race riots of 2001. The city has naturally been keen to shake off this latter image by embarking on a multi-million pound regeneration scheme.

The city's social and economic base has completely changed following the collapse of textile manufacturing. The symptoms of social deprivation are all too evident. Parts of Bradford are among the poorest areas in England, being sixth worst for unemployment (4.6% compared with the national average of 2.8%) and fifth worst for low incomes (44% of children live in low-income households compared with the national average of 27%). In 2000, only 34% of pupils achieved five or more GCSEs at A* to C grades compared with the national average of 50%. A large Muslim population plays a significant part in the city's business and cultural life. Immigrants from south Asia were attracted here in the 1970s by jobs in the textile industry. Pakistani and Bangladeshi communities make up a fifth of the city's population. A large proportion of the old housing stock is substandard.

Bradford is trying to erase its negative image and to exploit a number of its positive attributes: multiculturalism, services and tourism. Heritage tourism is being encouraged. Many factories of the Industrial Revolution have become museums, craft centres and galleries, or have been subdivided to provide small business units. Warehouse conversions are increasingly popular. Saltaire, famous as a Victorian model village created by Titus Salt for his factory workers, has been designated a UNESCO World Heritage site, giving it the same status as the Taj Mahal and the Pyramids of Egypt. A scheme called 'Vision of Bradford in 2020' has recently been implemented. It is based on a 'park in the city' concept that offers open green and leisure spaces, re-introduces water into the city centre and highlights much of the city's heritage of listed buildings. The scheme involves four regenerated quarters within the heart of the city. Each will create new spaces for commerce, education and leisure, as well as incorporating the natural feature of the River Aire, long buried under the city. However, in the recent credit crunch, this scheme seems to have been put on hold. There is also the National Museum of Photography, Film and Television and a range of annual festivals.

The issue is whether all this regeneration can really tackle the fundamental problems of Bradford's historical legacy. Regeneration is about more than bricks and mortar. The goals for any successful regeneration project are to breathe new life into a city and create an environment that supports the people who live and work there.

Southampton has a worldwide reputation as a port city. This fame was founded on the great ocean liners that carried huge numbers of passengers across the Atlantic and to and from the British colonies in Africa, Asia and Australasia during the nineteenth and first half of the twentieth centuries. In order to remain in business as a port, the city has had to make regular changes to its port traffic. When air transport reduced ocean-going passenger traffic, the port turned first to the cross-Channel ferry business and then to containerised cargo. Competition from nearby Portsmouth meant that there was little success with the passenger ferry business, but the container side has boomed. However, this is now under threat because the port has run out of waterfront space.

Like many other cities, Southampton has tried to regenerate its central area and maintain a strong retail centre. This has been successfully undertaken. This has involved three flagship developments: Ocean Village — a large marina development with housing and leisure facilities, Southampton Oceanographic Centre — a world-leading ocean research institution, and West Quay Retail Park — one of the largest projects of its kind in Europe. On the other side of the River Itchen, another flagship scheme, called Woolston Riverside, has just been set up. This development will be built on a 12-hectare brownfield site. It will transform a redundant shipbuilding area into a mixed-use waterfront area, with a marine business park, high-quality housing, and retail, leisure and community facilities.

In conclusion, it is clear that there comes a time in the history of most towns and cities when deliberate attempts have to be made to revive flagging fortunes. Following London's lead, Bradford and Southampton have both needed to invest heavily in regeneration — Bradford as a consequence of deindustrialisation and Southampton in order to boost its status as a leading port and regional shopping centre. It is clear that in each case there has been significant improvement in the social, economic and physical environment, yet at the same time issues remain.

36/40 marks awarded. This is a very good answer and, with a length of 1170 words, demonstrates many of the characteristics of a high quality response. The answer is well structured (being paragraphed in a logical manner) and purposeful. The student begins with a clear introduction which sets the scene for the essay to follow. Although the definition of regeneration is not given, it is clear from the whole response that the term is fully understood. The essay also ends with a conclusion which brings together many of the points raised in the body of the essay.

The examples are dealt with in chronological order, beginning with perhaps one of the best-known regeneration projects in the UK, the London Docklands, and then moving into more modern schemes in Bradford and Southampton. There is sound and frequent evidence of thorough, detailed and accurate knowledge of each of the schemes, and they are well developed. In addition, there is detailed understanding of the background to each of the schemes or areas. In the case of the London Docklands, the overall aims of the UDCs are given, whereas for both Bradford and Southampton, a background to the issues facing each of the cities is given. The section on Bradford is particularly good. Each of these sections demonstrates good critical understanding of the principles behind regeneration in these areas. The student has also recognised that while each of the schemes is in the UK, they are different in style and purpose.

The student has chosen to discuss the effectiveness of each of the schemes in turn. This is a valid approach. Evaluative statements are given for each of the schemes, but again in variable forms. For London, the student has given firm statements of success or otherwise; for the other two the statements are more speculative. This again is appropriate as the schemes have to be given time to succeed. To some extent this also provides evidence of critical understanding.

Overall, the answer demonstrates strong evidence of synthesis — recognising the complexity of regeneration in urban areas, in terms of its social, economic and environmental impact. In order to access Level 4, the student needs to demonstrate elements of flair, or insight, or maturity of understanding, and to write in a confident and coherent manner. Sometimes this is difficult to pinpoint. Here, one phrase exemplifies this and stands out: 'Regeneration is about more than bricks and mortar.' Based on the level of both breadth and depth of response, the answer accesses Level 4: 36 marks.

Development and globalisation

Question 9

(a) Study Figure 1 which shows the distribution of Toyota manufacturing plants outside Japan and the company's worldwide production in 2004.

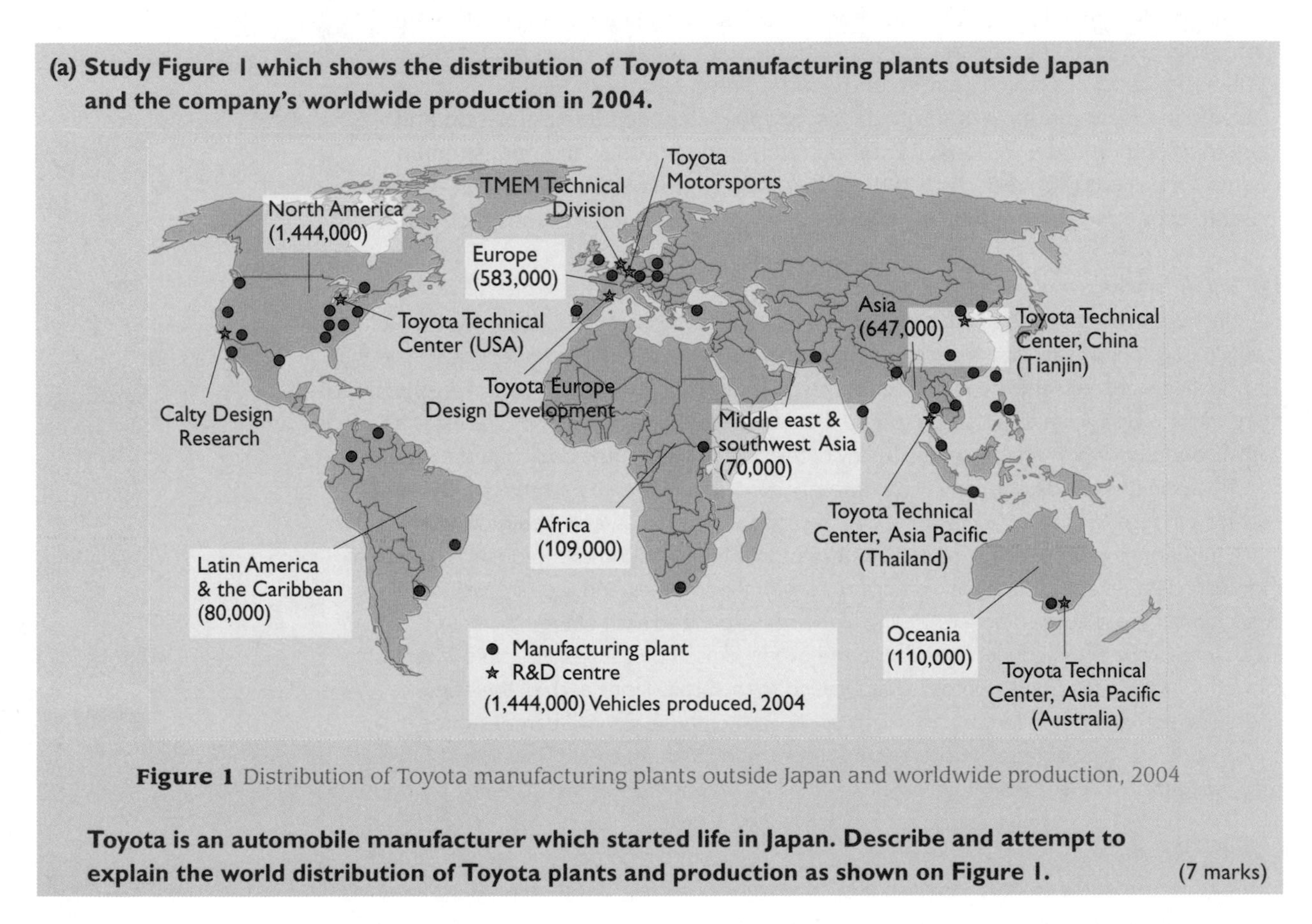

Figure 1 Distribution of Toyota manufacturing plants outside Japan and worldwide production, 2004

Toyota is an automobile manufacturer which started life in Japan. Describe and attempt to explain the world distribution of Toyota plants and production as shown on Figure 1. (7 marks)

e Note the use of two commands — you have to answer both elements of the question.

Mark scheme

Level 1: simple statements with regard to distribution. Explanations are little more than production in home country followed by a need for Toyota to produce where it sells (i.e. its markets). (1–4 marks)

Level 2: detailed statements with regard to distribution, particularly with regard to technical centres. Explains in detail why plants are scattered, such as reduced taxation and subsidies and grants. Sees peculiarities of local markets requiring specialised R&D. (5–7 marks)

Student answer

Toyota, outside of Japan, is found all over the world. Plants and R&D centres are found in every continent, particularly in North America, Europe and southeast Asia. There are only two plants in Africa and one plant and one R&D centre in Oceania. Toyota is a transnational company and wished to sell motor vehicles in markets around the world. In terms of cost, the company thought that it would be cheaper to produce cars where it was selling them. Therefore it established car manufacturing plants in Europe and North America, which have large production figures, and in other markets such as South America and Oceania.

e **4/7 marks awarded.** Simple statements are given with regard to distribution. R&D centres are mentioned but their distribution is not dealt with separately. The only explanation given is in regard to Toyota's markets and how production has expanded into those areas. There is no explanation of R&D. This fits the descriptor for Level 1 in the mark scheme. The description of the distribution is reasonable and although the answer only covers the need to manufacture in overseas markets, it is well done. The mark would therefore be at the top of this level and 4 marks out of the maximum 7 would be awarded.

(b) What can be the effects in the donor (home) country when transnational companies move their investment to other countries? (8 marks)

e Note the word 'effects' — i.e. more than one. You could consider social and economic effects, as well as short-term and longer-term effects.

Mark scheme

Level 1: straightforward statements referring to unemployment, poorer areas and on the positive side, more money. Answer mainly stresses the downside of the move in terms of the donor country. (1–4 marks)

Level 2: more detail on the material, particularly with reference to the downward spiral. Much more detail on the movement of capital back to the donor country. (5–8 marks)

Student answer

When TNCs move their operations from their country of origin to other countries, a lot of people may lose their jobs. This is not only true for the main company, but a lot of employment may also be lost in companies which supply components to the TNC. This is particularly the case in assembly industries such as motor vehicle manufacture. It also affects those firms which offer services to the TNC's factories, such as maintenance companies and even those which supply food to the factory's canteens. This movement can also affect service industries as has been seen with call centre operations of financial companies (banks, insurance companies) being moved to India.

All of this can have a major impact in regions where the TNC was located. There is less money to spend, fewer goods and services are required and the region can suffer increased unemployment. This is the reverse of the multiplier effect and is

often known as a downward spiral, or vicious circle. On the positive side, if the TNC is successful in its move, profits will be returned to the home country and paid out in the form of dividends or to the government as company taxation. As the headquarters of the TNC will probably stay in the home country, a successful company may need more high salary people at head office to run it.

e **7/8 marks awarded.** A wide range of points is made in this answer, many of which are outlined in a sophisticated manner demonstrating good understanding of the topic. For example, there is good detail on to the ways in which TNC movement can affect a region and create a vicious circle. As a balance, some information is also given on the positive sides of the move, more than just stating that money (profits) is returned. There is also some reference to the benefits of keeping the headquarters at home. This material fits the descriptor for Level 2 and such a comprehensive answer would be awarded 7 marks. With a little exemplification, the answer could have achieved the maximum.

(c) Discuss the role of transnational companies in the development of the global economy. (10 marks)

e The command 'discuss' requires you to present a verbal debate — some evaluation of 'role' is needed.

Mark scheme

Level 1: simple statements commenting on the importance of TNCs across the world without developing their role in the global economy. (1–4 marks)

Level 2: clear statements with a precise picture of how TNCs operate across the world and their importance in the global economy. Good use of case study material to exemplify points. (5–8 marks)

Level 3: clear discussion of the global economy and of the role of TNCs in it. Good and accurate use of exemplar material. (9–10 marks)

Student answer

During the twentieth century a new global economy has emerged through a process known as globalisation. This is defined as the increased interconnection between the world's economic, cultural and political systems. In the global economy, transnational corporations (TNCs) and nation states are two of the major players. The growth of TNCs has been particularly rapid in the second half of the century, because in the 1960s less than 10,000 companies were TNCs, whereas today some economists have estimated that there are over 60,000 in operation.

TNCs have been the major force in increasing economic interdependence between various parts of the world. TNCs initially emerged from developed countries but increasingly they are the product of economic growth in a number of countries which are known as Newly Industrialising Countries (NICs), e.g. South Korea and Singapore. TNCs have their headquarters in their country of origin, along with R&D, but their manufacturing plant will be overseas, probably in a cheaper labour location.

TNCs have stimulated world trade as they move goods and materials between their factories. This is particularly the case for those industries involved in assembly, such as the motor vehicle industry. TNC involvement in poorer countries such as South Korea enabled such countries to become richer through the benefits of employment and the transfer of technology and management skills. This is why several large firms developed in South Korea which eventually became TNCs by locating factories overseas, e.g. Samsung, Hyundai and LG.

In the early 1950s, 95% of all manufacturing was concentrated in the industrial economies. Since then, TNCs, as a result of foreign direct investment (FDI), have been involved in a movement of manufacturing to developing countries, a process known as global shift. The creation of TNCs is not confined to resource exploitation and manufacturing as many service companies have now become global in their operations. The accountancy industry is a very good example with the emergence of four very large corporations, e.g. KPMG. One recent trend, which has affected the global economy, is the growth of services such as call centres, where large banking and insurance companies move their call centre operations from Europe and the USA to such countries as India. Some of these movements have resulted in deindustrialisation in developed countries and decreased employment opportunities in developed countries.

e **10/10 marks awarded.** This is an excellent answer. TNC structures are described and their movements explained. The emergence of a new global economy is well covered, particularly the growth of NICs and the development of TNCs within them as a result of foreign direct investment. The global economy and TNCs are brought together in a discussion of the role of such companies in bringing about the present world economic order. Full use is made of exemplar material, which is both accurate and well used. Such material fits very well into the descriptor of Level 3 in the mark scheme. With a very good understanding shown of both the global economy and transnational corporations, along with good use of exemplar material, such an answer would be awarded a mark at the top of the level, probably the maximum 10.

Question 10 **Essay question**

Discuss the varying roles of (a) the promotion of trade and (b) the provision of aid, as approaches in the efforts to raise living standards in the poorest countries of the world. (40 marks)

e The command 'discuss' requires you to present a verbal debate — some evaluation of 'role' is needed.

Appropriate content for a response to this question might include:

- the problems of the poorest countries which need to be addressed
- the benefits of trade such as increases in the amount of wealth being generated, allowing an increase in living standards
- the belief that countries should go through a process of industrialisation (just like those in the 'North'). This would allow more trade, increasing the revenue flowing into the country

- the doubts, expressed by some economists, that many of these countries cannot be competitive as they have too many problems such as AIDS, internal conflicts and climatic problems, e.g. drought
- the main systems of international aid and how they work
- the argument that aid does not always get to where it is needed and it is often not used effectively
- aid dependency and aid with 'strings'

Synopticity could emerge with some of the following:

- a critical understanding of the extent of the problems experienced by developing countries (economic and social) and how it is possible to resolve them in general terms
- a critical understanding of how trade works in a global context
- a critical review of the arguments for and against trade as a solution to problems of developing countries, particularly considering the factors on which success will be based (adoption of capitalism, the trickling down of wealth, promotion of free trade)
- an understanding of the provision of aid
- a critical review of aid, involving both the donors and recipients
- evidence of breadth/depth of exemplar material

The question clearly requires a discussion and the response should try to come to a view between the effectiveness of the two approaches. Any conclusion can be credited as long as it is measured and reasonable, and related to the content of the answer.

Student answer

For the last 40 years or so, the United Nations has recognised the poorest countries in the world by placing them into a separate category known as least developed countries (LDC). These are countries where there is a very low GDP and low levels of nutrition, health provision, education and literacy. Many people exist on an income of less than $1 per day, so it is no surprise that the quality of life of most of the inhabitants of such countries is extremely low. Many of the countries are in serious debt as a result of borrowing money from Western banks and other financial institutions. There is no hope that most of this debt will be repaid and all the time, these countries have to try to pay the interest on their borrowings. Some of the policies that have been proposed to try to lift such countries to a higher stage of development have largely centred upon trade or aid.

Trade can stimulate economic growth within a country by generating more wealth, this, in turn, helping to raise the living standards of the people. Some economists believed that the way forward for the world's poorer countries was to industrialise along the lines of those countries in the West who had done it before. This would provide more revenue from manufactured goods than had been the case from the raw materials and agricultural products that had been the main forms of exports in the past. To do this, such countries had to embrace Western-style capitalism and to allow the benefits to trickle down to all levels of society. On a global scale, free trade had to exist, or certainly trade had to come with as few restrictions as possible.

Other economists doubt whether trade can bring about such changes within these countries. They point out that these countries have to contend with internal wars, widespread AIDS and corruption within government and they also doubt whether any wealth generated can trickle down to benefit all but a small minority of the population. The debt burden which they suffer puts them in a difficult position, because in order to receive help they have to submit to conditions laid down by the IMF and the World Bank, which may lead to cuts in health, education and welfare programmes.

Aid can be supplied to LDCs through various agencies including governments, international organisations (IMF, UNESCO) and charities such as Oxfam and Comic Relief. Aid can be in the form of money, but it can also consist of goods, technical assistance, development projects and education programmes. When disasters strike LDCs (hurricanes, earthquakes, volcanic activity) it is important that short-term relief is sent to them. Long-term aid is usually in the form of some project for which capital, equipment and technical knowledge will be supplied. Projects could take the form of building a dam along with HEP and irrigation provision, or a programme to improve the quality of teaching in rural schools. Many economists believe that the way in which the aid is delivered is very important. Some schemes are 'top-down' where a responsible body (internally or externally) will direct operations, or 'bottom-up' where the supplier of the aid works closely with local people and uses local ideas and knowledge. Such schemes are usually funded by NGOs such as Oxfam. Many economists see such schemes as a means of developing a country, but there are many critics. It is often alleged that most aid does not reach the people it was designed for, and even when most of it gets through, critics say that is not always used effectively and that corruption is one of the obstacles. As most of these countries possess a poor infrastructure, this is another reason why aid cannot be used effectively as it is often difficult to reach some of the remotest parts. Some aid may also come with strings attached, i.e. restrictions imposed by the donor and many worry that people within LDCs will become too dependent upon it and not have a great deal of incentive to improve themselves by their own efforts.

Aid can also come from private contributions and philanthropy such as the Bill and Melinda Gates Foundation set up by the founder of Microsoft. Celebrity philanthropy is also being led by people such as Bob Geldof and Bono such as Live Aid in the past.

At the 2005 Gleneagles G8 meeting it was agreed to increase Overseas Development Aid (ODA) to 0.7% of GNP yet, in 2006 and 2007, it receded by 15%. This has happened after every major international conference. The percentage of the GNP of developed countries allocated to ODA fell in 1993 after the Rio Summit in 1992, and again after the Millennium Summit in 2000. The developed countries commit solemnly to increase ODA and proceed immediately to reduce it. Sir Bob Geldof has been highly critical of this. The Monterrey Conference recognised that aid effectiveness was being compromised by burdensome donor-imposed conditions such as the tying of aid, and by lack of harmony in the operational procedures of donors, including multilateral institutions such as the World Bank.

It can therefore be seen that both aid and trade policies come with their problems and historically have been limited in what they have accomplished. This is not to say that the use of either approach has been an outright failure, as there are a number of success stories, but it is also true to say that there have been many failures of such policies. Trade and aid policies by themselves are not enough to make a significant difference to LDCs, but they do have a role to play in the much greater framework of international development.

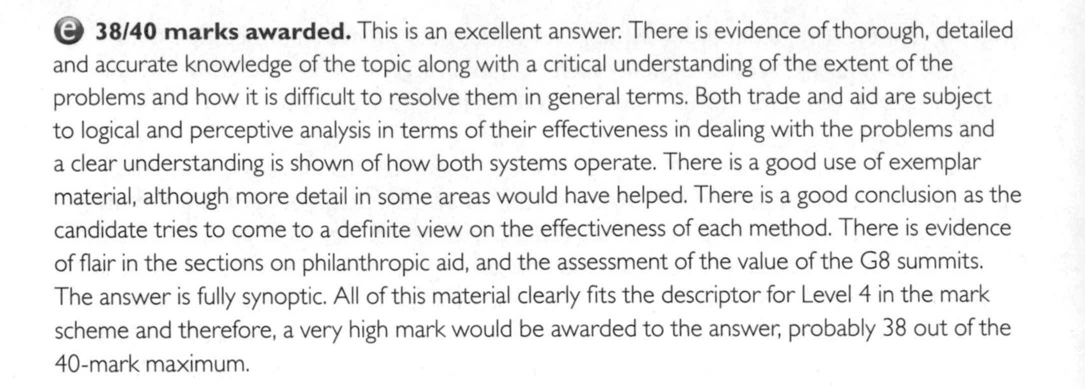

38/40 marks awarded. This is an excellent answer. There is evidence of thorough, detailed and accurate knowledge of the topic along with a critical understanding of the extent of the problems and how it is difficult to resolve them in general terms. Both trade and aid are subject to logical and perceptive analysis in terms of their effectiveness in dealing with the problems and a clear understanding is shown of how both systems operate. There is a good use of exemplar material, although more detail in some areas would have helped. There is a good conclusion as the candidate tries to come to a definite view on the effectiveness of each method. There is evidence of flair in the sections on philanthropic aid, and the assessment of the value of the G8 summits. The answer is fully synoptic. All of this material clearly fits the descriptor for Level 4 in the mark scheme and therefore, a very high mark would be awarded to the answer, probably 38 out of the 40-mark maximum.

Contemporary conflicts and challenges

Question 11

(a) Study Figure 1, which shows the percentage of people living in Great Britain who were born in the UK by ethnic group (2001).

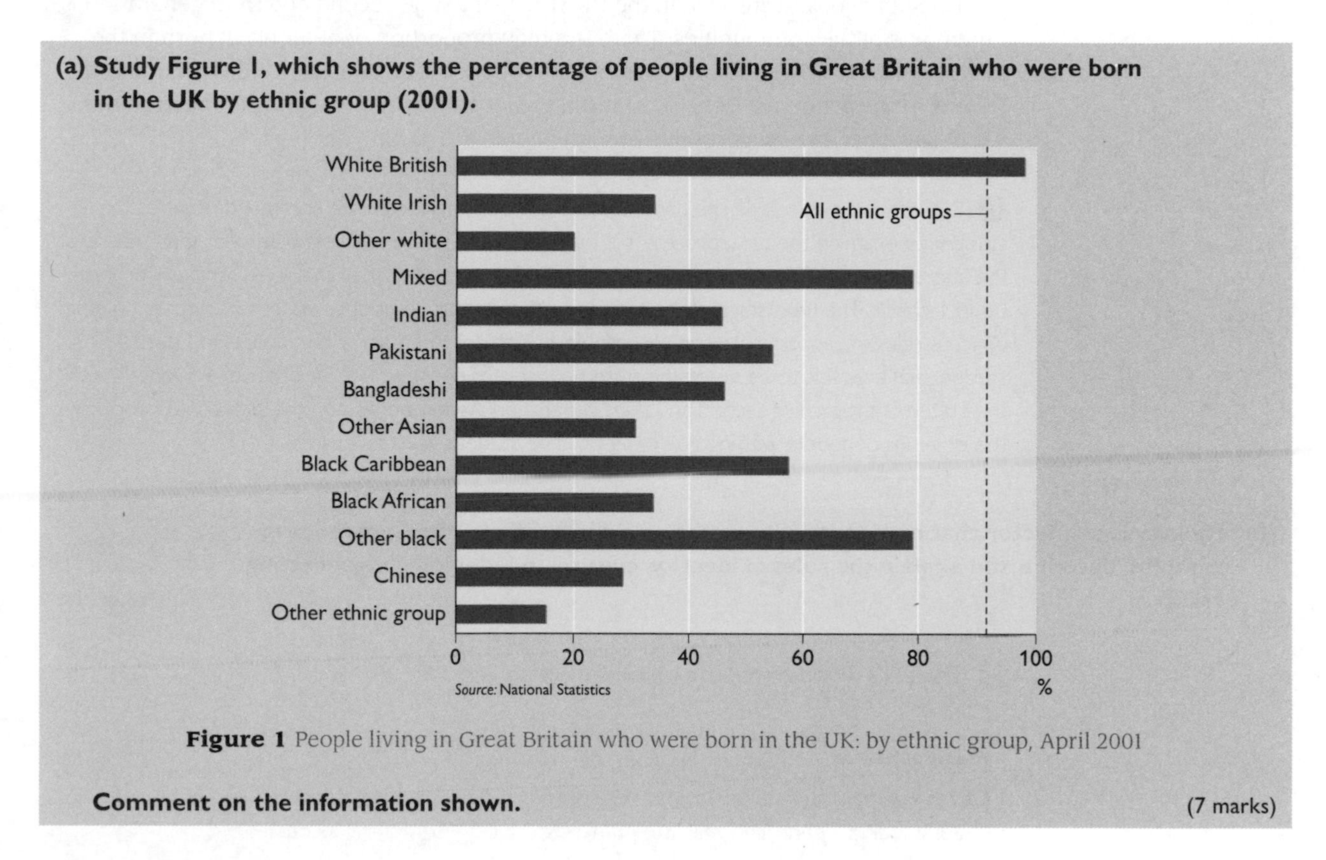

Figure 1 People living in Great Britain who were born in the UK: by ethnic group, April 2001

Comment on the information shown. (7 marks)

Note the command word 'comment on'. You are required to make an observation or inference that arises from the data.

Mark scheme

Level 1: simple 'lifting' of figures from the diagram with the use of high/low statements to illustrate basic understanding. No commentary that is meaningful. (1–4 marks)

Level 2: more sophisticated description of the data, with perhaps some classification. Valid and appropriate commentary that arises from the data. (5–7 marks)

Student answer

The majority of white British were born in the UK and are therefore indigenous to the UK population. This is above the national average as you would expect. If you look at the next set of ethnic groups underneath (Irish white and other white) very few of these were born in the UK. These could be Australians, or French, as well as Irish, who have migrated to the UK for jobs and will probably not settle here. It is interesting to note that over 40% of Indians, Pakistani, Bangladeshi and black Caribbean people were born in the UK. These must be second and third generation members of the communities. There is a high proportion of other black born in the UK. These must be South Americans. Thirty per cent of Chinese were born in the UK. Some of these figures seem to contradict what some right-wing parties such as the BNP say when they suggest that foreign people are 'taking our jobs'.

e **7/7 marks awarded.** The command here is 'comment on'. This command requires the student to examine the data, process it in his/her mind and then make a comment that arises from the stimulus material. In many cases students will describe what they can see — this will achieve Level 1 credit. The next stage of the task is to process that descriptive statement into something which is relevant, appropriate and geographical. Such commentary allows access to Level 2. This student achieves this three times: the statement regarding white people other than white British; the statement regarding second and third generation Asian people; and the statement concerning the opinions of people with right-wing views. The student gains full marks.

(b) Ethnicity is one factor that may contribute to the origin of conflict. With reference to examples, describe and explain the roles of identity, culture and ideology in generating conflict. (8 marks)

e There is a requirement for 'examples'; therefore 'breadth' is essential.

Mark scheme

Level 1: simplistic statements which are generalised and lacking in depth of understanding. There is no real attempt to introduce geographical examples of where different reasons for conflict exist. (1–4 marks)

Level 2: more sophisticated comments on the origins of conflict (covering at least two of the factors). There is a clear attempt to support the argument with references to locations and places. (5–8 marks)

Student answer

Identity is a sense of belonging to a group or geographical area where there is some common characteristic, which could be language or religion. Differences in identity have shown themselves in the area once governed by the former Yugoslavia. Here there is a mix of many nationalities, Slovenes, Croats, Serbs, Bosnians, Albanians and Hungarians as well as a complex mix of Orthodox Christians and Roman Catholics and Muslims all due to complicated patterns of migration over the last

2,000 years. Each of these people have a strong feeling of identity, and that has been one of the causes of conflict in that area, especially after the fall of the communist government of Tito. At a much different scale, people in Lancashire and Yorkshire in the UK have a strong sense of identity, though it does not cause conflict these days (there was the former Wars of the Roses though in medieval times).

Culture refers to the customs and beliefs of a population, and their shared attitudes. Differences in culture can also be similar to differences in religion, and so many of the points made above regarding the former Yugoslavia could be made here. Many people in the world have found aspects of the culture of immigrants different, which has made them want to resist such people. This is pure discrimination, but is caused by differences in culture. This is probably the reason why many people don't like to have Romany people living nearby as they have a different culture. There was a recent example of conflict in the village of Seamer near Scarborough where the local people did not want a gypsy camp set up nearby and they marched through the camp, and the police became involved.

Differences in ideology have created conflict in Afghanistan, where the Taliban and US/European forces are fighting. Here there are strong differences in views over women's rights for example, and the education of girls.

e **8/8 marks awarded.** This is a good answer. The student understands the differences between the terms identity, culture and ideology and is able to exemplify each of their influences in generating conflict well. The opening paragraph is very strong and detailed, and even the reference to Lancashire/Yorkshire divisions reinforces understanding.

The second and third paragraphs are not as strong, yet they are clear, and there is a good 'sense of place' to them. The student has used examples briefly to support the point being made, and each one is appropriate and relevant. What has also made this answer very good, is that, probably without design, the student has touched on a range of scales of conflict: local (Seamer), regional (Balkans) and national/international (Afghanistan) — i.e. breadth. The response accesses Level 2: 8 marks.

(c) With reference to one recent major international conflict, discuss the impact of the conflict on the environment of the area affected. (10 marks)

e Note the key word in the question — environment.

Mark scheme

Level 1: simple statements of impact, which are generalised and non-specific to identified conflict e.g. land is mined, farming land is destroyed. (1–4 marks)

Level 2: detailed statements of impact with a clear sense of place being generated. The answer is detailed and makes sophisticated comments on impact. (5–8 marks)

Level 3: a fully developed answer, with good elaboration and clear and appropriate detail. Recognition of the complexity of the issue. Recognition of the changing impact over time. (9–10 marks)

Student answer

A war has been waged in the Darfur region of Africa for several years now. Darfur is a semi-arid province to the west of Sudan. Darfur alone is the size of France. Though Sudan is an Arab-dominated country Darfur's population is mostly black African. For years there have been tensions between the mostly African farmers and the mostly Arab herders, who have competed for land.

Now the problem is that the Arab Janjaweed have been accused of riding into the villages of the black African people and of killing the men, raping the women and setting fire to the villages. As a result the women and their children have left the villages and their land in order to go to somewhere that is safe. They have settled near the main towns of the area, where food, water and medical help are being provided by charities such as Medicins sans Frontières in huge tented settlements.

The environmental costs are many. The former villages now stand idle, many in burnt ruins. The plots of land around them are untended and they are reverting back to some form of natural vegetation. The area is on the edge of the Sahara desert and drought-resistant plants are being re-established. Many of these plants have long roots, and once established are very difficult to remove again. The former farm land is going to ruin too. This will please the Janjaweed as their people are nomads and want the land so that they can wander around with their animals.

Around the refugee camps, the people continue to try and find fuelwood for cooking and heating. Many women are too frightened to venture too far for fear of being raped. But, they do, often at night. Areas of woodland are disappearing fast. The same sort of problem continues to arise — the women in the camps and the towns themselves don't want to farm, and the land is deteriorating fast.

e 9/10 marks awarded. The key discriminator for Level 2 is the concept that the students must provide a sense of place to their answer, assuming that he/she has given detailed material that is relevant to the question. This candidate 'sets the scene' very well at the start of the answer, and the sense of place from then on is established. The answer is detailed in that the impacts of the conflict in two different types of area within the country are described, and explained. The reader is made aware of the environmental costs as required. The student has also made a number of sophisticated comments on the subsequent impacts — for example, the statement regarding drought-resistant plants becoming re-established. The student is demonstrating clear understanding of the task, and good knowledge of the environmental consequences of the conflict. The final paragraph also satisfies the requirement for Level 3 — the movement of women and children to other parts of the country is causing a knock-on effect elsewhere. This indicates a recognition of the complexity of the issue. 9 marks are awarded. How could this have been improved? Perhaps the student could have developed the impact on the human environment better — at the moment it concentrates purely on the physical environment.

Question 12 **Essay question**

With reference to examples, discuss the nature of, and reasons for, separatist pressures around the world. (40 marks)

e An essay of this type requires both breadth and depth for high marks.

Appropriate content for a response to this question should include:

- a description of the variety of separatist pressures around the world
- an outline of the reasons causing separatist pressures around the world
- recognition and discussion of the variety of aspects of/reasons for separatism in different parts of the world

Synopticity could be achieved by:

- understanding the context of varying timescales
- evidence in the breadth/depth of case study material
- detailed critical understanding of both the nature and causes of separatism
- detailed analysis of both nature and causes and a recognition that they may vary around the world

Student answer

Separatist pressures are tensions that occur when one or more regions of a particular country no longer want to be part of that country and wish to form their own nation. They may have a lack of autonomy which they campaign to obtain by both peaceful (democratic) and violent means.

A key example of an area where separatist pressures occur is in the Basque region of northeast Spain and southwest France. Here, four regions of Spain and three regions of France wish to separate from their current countries and join together to form their own country but so far they have been unable to do so. Despite being given greater autonomy in 1936 by the Spanish republican government this decision was reversed after the civil war of 1937 when General Franco came to power. Indeed the Basques were persecuted and thousands were forced into exile or executed before Franco's death in 1975. The language of Euskara was banned and the culture oppressed leading to the formation of a Basque nationalist group ETA in 1959. They declared war on Spain and attacked military and government officials and buildings and are still in existence today. More democratic motions have been the formation of the Basque Nationalist Party, and Herri Batasuna political party, although the association of the latter with ETA has led to imprisonment of its members and it was banned for 3 years from 2002 due to its affiliation with terrorist organisations. Today, the Basques enjoy some autonomy and control their own health, education, police and transport, but this is not enough for some.

A key reason for separatist pressures is that a region that wishes to separate has a history of independence or greater autonomy in the past, as was the case with the Basques. A region wishing to separate usually also has a different language to the rest of the country, e.g. Euskara is the language of the Basques, as well as a different culture. Often the region also has a different religion to the rest of the country, as is the case with the Basques where the Basque church is a different form of Christianity

to the Catholicism practised in the rest of Spain and France. The region usually has a geographically remote location too (the territory of the Basques is hundreds of kilometres from both the French capital Paris, and the Spanish capital Madrid). Finally separatist pressures may occur in a region which feels its resources are being unfairly exploited by central government — the Basques are responsible for a very large proportion of Spain's iron and steel production. Bilbao is the largest city in the Basque region, and it is a major port and centre for iron and steel production. Just to the east lies Guernica, a town famous for being bombed by Fascist forces in the civil war. Many Basques have never forgiven the Spanish government for this.

The increase in separatist pressures also has an important impact but these tend to be on a small scale. For example, a key effect of the separation of the Basque territory from Spain would be a damaging impact on the Spanish economy. The port of Bilbao among other cities in this region is responsible for around 95% of Spain's steel industry as well as an important fraction of the Spanish shipbuilding industry. The loss of this region with its resources and skilled workers would be devastating to the Spanish economy and would cause irreversible damage to these key industries. As a result the Spanish are keen to hold on to this region. In addition separatist pressures are important as they actually lead to a loss of life. With military organisations such as ETA carrying out terrorist actions such as bombings, people's lives can be at risk even when they have nothing to do with the separatism. Furthermore separatist pressures can lead to impacts in areas which are unconcerned by such activities such as the shooting of two Spanish police officers in the French town of Capbreton in December 2007 which isn't in the Basque region.

Around the world some other people are wanting to separate from their country. This is true in Scotland where many Scottish people want to separate. This is because they feel peripheral from the capital of the UK (London) and don't feel their needs are being catered for. Also many Scottish believe that their native language of Gaelic can only be preserved if Scotland becomes independent. One of the major reasons for Scottish independence was during the 1970s when North Sea oil was found and used. Many Scottish did not feel that they got what they should have out of that resource hence the SNP (Scottish National Party) began the 'It's Scotland's oil' campaign.

A consequence of these pressures is that Scotland now has its own devolved government, and can and does make its own decisions. For example, tuition fees at Scottish universities don't exist although English students going there have to pay. Some may say this is unfair, especially with fees going up to £9000 in England. The Scottish Nationalists are now the main party in power and are keen to push for full independence.

Another example of separatism is Chechnya where a group of people known as the Chechen rebels want to be independent from Russia. This is due to many reasons — one being historical as in the Second World War Stalin accused Chechnya of collaborating with the Germans and many were sent off to Siberia as a punishment. It is thought that 800,000 people were sent and unable to return till the mid 1950s. Religion can also be a reason for separatism in this case. Chechnya is a majority Muslim country whereas Russia is a Russian Orthodox state. The rebels were also infuriated that when the Russian state collapsed neighbouring countries such as Georgia were granted independence whereas Chechnya was not.

So, it can be seen that a variety of pressures for separatism exist around the world, and each of the areas where they exist have a different combination of reasons for them.

e **26/40 marks awarded.** This answer is a solid Level 3 response, at just over 1000 words. The student answers the question as set — a variety of examples are referred to, and the response deals with both the nature of, and reasons for, separatism. The strength of the answer lies in the variety of reasons, each well supported by the use of case studies examined in depth. The Basque case study is more detailed, and does venture more into the historical background of the conflict (though it is true that historical factors are important for this region), and this part of the answer is repetitive at times.

There is frequent evidence of thorough, detailed and accurate knowledge and understanding, and the explanations are certainly direct and logical. Each of the examples discussed: the Basques, Scotland and Chechnya are developed. There is some evidence of synopticity.

The main weakness within this level of credit is the lack of critical understanding of the nature of, and reasons for, separatism. Only occasionally does the student attempt to assess the strength of the demands for separatism, or evaluate the effectiveness of those demands. There are suggestions of this in the section on the Basques (for example, the sentence concerning the shooting of the two Spanish police officers in Capbreton), and in the section on Scotland where more could have been made of the Scottish Parliament's ability to make decisions vis-à-vis England. Overall therefore the response is purposeful and well structured. It is well within Level 3, and would gain the mid mark — 26 marks.

Knowledge check answers

1 The evidence is that new land is continually forming under the oceans where plates are moving apart, allowing magma to come to the surface, and the evidence obtained from palaeomagnetic surveys shows that this has been going on for millions of years. Sea floor is also being destroyed where plates converge. The continuing presence of earthquakes and volcanic activity is also evidence of the ongoing nature of these processes.

2 The movement of oceanic crustal plates away from constructive (divergent) plate margins is brought about by convection currents within the asthenosphere. As the plates diverge, magma is drawn to the surface, giving basaltic volcanic eruptions that form into mid-oceanic ridges. Under the Atlantic Ocean, evidence shows that the sea floor becomes older with distance from the mid-Atlantic ridge.

3 Subduction occurs where convection currents in the asthenosphere cause oceanic plates to move downwards away from the surface at destructive (convergent) plate margins. Surface features that result from subduction include oceanic trenches (the point where the plate is pulled downwards). As the plate is driven deeper, the temperature rises (in the Benioff zone), melting the descending plate, and the molten material thus created attempts to find its way back to the surface. This rising magma results in volcanoes that are often explosive, forming a line known as an island arc when two oceanic plates are involved in subduction. A subducting oceanic plate will cause the opposing continental plate to rise, with the marginal sediments being forced up into fold mountain ranges.

4 Hot spots are stationary and a volcano will form over them. When a line of volcanoes is seen (such as in the Hawaiian Islands) where only one is active and the remainder extinct, the plate must have moved, the extinct volcanoes marking the previous positions of the plate over the hot spot.

5 You first need to make a list of the types of area where volcanic activity is present. These should be: mid-oceanic ridges (such as the mid-Atlantic ridge), major rift valleys (East African), subduction zones (such as the Peru-Chile oceanic trench off the western coast of South America), and hot spots (the Hawaiian islands are the best example). Now locate these on a world map.

6 When plates move, there is interaction along their edges. This causes earth movements resulting in earthquakes and allows molten material to escape to the surface during volcanic activity. Not all plate boundaries have both volcanic activity and earthquakes although most do (such as western South America and other parts of the Pacific Ring of Fire, and along the Mediterranean Sea).

7 The impact of a tsunami depends upon the event that generates it, the distance travelled, the height of the waves, the coastal geography and density of population, and the possibility of issuing warnings to the coastal population.

8 It is impossible because of the forces that are being unleashed when these events happen. It has been suggested in California that it might be possible to 'lubricate' the plate movement to stop the sticking (and then release), which is why earthquakes occur there. Because there is no possible prevention, people have to concentrate on methods of prediction and protection to keep themselves safe. With earthquakes, as there is no reliable method yet devised for prediction, all the focus is on making sure that you survive the event.

9 These are latitude (distance from the equator), altitude (height of the land), continentality (distance from the sea), prevailing winds, seasons and ocean currents.

10 Generally speaking aridity increases with distance inland so parts of the interiors of large continents, such as Asia, feature deserts (e.g. the Gobi desert).

11 Weather is the day to day variation in temperature, precipitation, winds, pressure, sunshine hours and humidity, whereas climate is the average monthly pattern of these measured over a period of some 30 years.

12 Heavy snow in winter is usually associated with either an Arctic Maritime or occasionally a Polar Continental air mass when moisture is picked up as the air passes over the North Sea, whereas a heatwave in summer will be the result of a Tropical Continental air mass over the British Isles.

13 In winter the impacts relate mainly to the cold and fog. Energy generation peaks so the costs and problems of supply are important — with related health issues — in particular, fuel poverty, hypothermia and the elderly. Traffic issues might result from fog.

In summer dealing with drought is paramount, both on the domestic and agricultural front. Heatwaves can also cause spontaneous fires, which are particularly a problem in rural areas, such as the National Parks. Additionally, heatstroke affects the most vulnerable, and is most likely in urban areas where temperatures are further inflated by the heat island effect. Finally, air quality deteriorates, particularly in cities, where vehicle emissions react with sunlight to cause photo-chemical smog.

14 Responses to storm events tend to be reactive and short term, whereas it is easier to prepare for drought associated with high pressure as such conditions occur over a longer time period (so responses can be proactive to a certain extent). Storm conditions result in traffic management (e.g. the Severn Bridge is closed to traffic and Operation Stack takes place on the M20 when the channel ports are closed to shipping). In contrast, drought management plans are put in place by the water companies and local authorities, and in extreme circumstances water is rationed.

15 Diurnal range is the temperature difference, usually measured in degrees centigrade, between the highest and lowest temperature recorded in one day, whereas the annual range is the difference between the lowest monthly average temperature and the highest monthly average over 1 year.

16 The ITCZ is located around the Earth on average between 5° north and 5° south of the Equator, where the northeast and southeast trade winds converge in a low-pressure zone known as the inter-tropical convergence zone. The location of the ITCZ varies throughout the year, moving into the northern hemisphere to follow the sun between March and September and into the southern hemisphere for the rest of the year. As it moves it takes with it the zone of low pressure, which is associated with seasonally heavy precipitation.

17 The urban heat island refers to the tendency for a city or town to remain warmer than its surrounding rural area by as much as 10 degrees centigrade. The Venturi effect occurs in large cities where wind is forced between buildings. When the wind blows through a street between tall buildings the pressure at the two ends of the road will differ, forcing the moving air to the 'low pressure' side.

18 As distance from the city increases so does the quality of the air. This relationship is also generally true for relative humidity, sunlight and wind speed.

19 Many scientists have looked at the historical trace of climate change and suggest that the Earth has naturally fluctuated between hot and cold periods and does so approximately every 1,500 years. In addition to periodic changes in energy from the Sun itself, the Earth's orbit also varies slightly, thereby bringing us closer and further away from the Sun in predictable cycles (called Milankovitch cycles). Evidence for past climate change can also be seen by scientists studying, among other things, tree rings (dendrochronology), fossils and oxygen isotopes in ice.

20 This is the natural process where the atmosphere traps some of the Sun's energy, warming the Earth enough to support life. It occurs as a result of heat absorption by greenhouse gases in the atmosphere and re-radiation downward of some of that heat. Water vapour is the most abundant greenhouse gas, followed by carbon dioxide and methane.

21 A biome is global in scale whereas an ecosystem can be any size.

22 The biotic components are the living parts of an ecosystem whereas the abiotic components are non-living.

23 Energy is lost through life processes, such as respiration, movement, excretion and reproduction; roughly 10% of energy is passed on to the next trophic level.

24 A Gersmehl diagram shows how nutrients are recycled within an ecosystem. It identifies the inputs, outputs, transfers and stores of nutrients and gives a visual impression of the relative importance of the size of the stores and transfers.

25 A prisere is another name for a plant succession and is used when the pioneer community colonises bare ground. The climatic climax is the final stage in a plant succession.

26 This has the second highest NPP (net primary productivity) of all biomes, after the tropical rainforests. Although the vegetation lies dormant during the winter growing conditions are favourable for more than half of the year.

27 A secondary succession is one that develops again through a series of seres after a previous interruption, e.g. following drought or on a derelict building site, whereas a plagioclimax is a plant community permanently influenced (arrested) by human activity.

28 De-industrialisation resulted in many derelict factory sites, canals, railway sidings and mines, which, left untended, were slowly covered in moss, weeds and then brambles, eventually becoming overgrown scrub as a secondary succession took place.

29 The verges of routeways, usually vegetated, provide a good home for wildlife as they are generally undisturbed by people. Moving traffic and freight can transport and deposit exotic seeds from other ecosystems. Sometimes invasive plants can overtake and out-compete indigenous ones.

30 A fragile environment is one that is delicately balanced and easily disturbed. Once disrupted it is extremely difficult for a fragile environment to recover. Sustainable management allows an environment to be used for the benefit of the current population but ensures that in doing so it is not damaged in any way for future generations.

31 A megacity is a city with over 10 million people. There were 28 of them in 2010. The majority of these (18) were in the developing world. An example would be Istanbul or Shanghai. A world city is one that is connected commercially and technologically with similar cities in the world, and together they are globally powerful and have a huge influence in terms of trade, political strength and innovation. Examples include London, New York and Tokyo.

32 Urbanisation is the increase in the proportion of a country's population that lives in towns and cities. Suburbanisation is the movement of people from living in the inner parts of a city to living on the outer edges, or in areas between radial routes from the centre. Counter-urbanisation is the movement of people from large urban areas into smaller urban areas or into rural areas, leapfrogging the rural–urban fringe.

33 Re-urbanisation is the process whereby areas of large towns and cities that have been experiencing a loss of population are beginning to slow this decline, or even reverse it and begin to grow again. It represents a movement of population towards the central city area with a rise in the number of housing developments. It is associated with regeneration, often in the form of city centre apartments and waterside developments, and increased employment in the financial and service sectors. Regeneration involves social, economic and environmental renewal of a previously rundown urban area.

34 A wide number of case studies exist. Examples include:
- gentrification — Notting Hill and Islington in London; Jericho in Oxford
- property-led regeneration —London Docklands, Salford Quays
- partnership — Hulme, Manchester; Greenwich Millennium Village

35 Reasons for out-of-town retailing areas:
- increased access to a wealthier suburban population
- cheaper land for site and expansion
- access to new roads/motorways
- advantage of derelict land grants (brownfield sites)
- high levels of dereliction in inner city/CBDs
- difficulty of access to CBDs — for shoppers and deliveries
- possibility of developing new attractive greenfield sites

36 For example the Trafford Centre:
- Highly accessible — located adjacent to junctions 9/10 of the M60 with convenient access to the M602 and an excellent dual carriageway link to the city centre.
- 10,000 free parking spaces. Car users entering the site encounter a vehicle messaging system that communicates messages about car park availability on site.
- Latest additions include Barton Square, a unique home retailing section including Next Home and Legoland.

37 For example Doncaster:
- A redeveloped indoor shopping centre called the Frenchgate centre with an anchor store of Debenhams.
- A new bus interchange, integral to the Frenchgate centre, where all loading bays are inside a pollution-free environment.
- Pedestrianisation of the area around the old Market Hall, which has been renovated.

38 The assumption here is that you have studied the waste management of one urban area — possibly the area where you live. All urban areas in the UK have to meet EU targets for recycling. Be able to provide details of how this urban area tries to recycle its waste, for example the number and location of household waste disposal sites, or even the ways in which it sorts out recyclable materials.

39 Problems caused include the following:
- A high proportion of the population work in urban areas but live in rural or suburban areas causing problems associated with commuter traffic.
- Many journeys to work are around and across cities — public transport is often not developed for this.
- Economic growth has led to more commercial vehicles on the roads.
- People often use their vehicles to access leisure and education facilities (e.g. the 'school run').
- Atmospheric pollution can lead to bronchial complaints and eye irritation. Asthma is an increasing problem, possibly exacerbated by traffic fumes. Concentrations of low-level ozone are increasing, which can lead to photochemical smog.

40 A wide range of transport management systems exist that could be studied: new road schemes (e.g. the M25), restricted access schemes (e.g. the Congestion Charge), traffic management schemes (e.g. park and ride, bus lanes), mass transit systems (e.g. the Metrolink). Make sure you give precise details of your chosen system.

41 An increase in a country's level of wealth will bring about a greater investment in medical provision. This will reduce infant mortality, which after some time should result in a falling birth rate as contraception becomes more widely available and the benefits of smaller families can be seen. Greater medical provision also leads to an increased life expectancy. The quality of life for the population improves with better education (higher literacy rates), sanitation and housing, all of which are paid for through greater economic development.

42 You need to make lists covering developed countries, developing countries, LDCs, NICs, RICs, centrally planned economies and oil-rich states. Try to fit them into the development continuum.

43 Think of when large-scale manufacturing began (in the nineteenth century) and then go on to look at the global distribution of manufacturing today. The terms that are picked out in the text in bold type should help you with the answer.

44 Foreign companies wished to take advantage of the huge labour force available whose cost would be well below the equivalent in many other countries. China was also moving from a centrally planned economy to a more market-orientated system, which created a huge market for consumer goods, and the government had set up special economic zones (SEZ) where foreign companies could get favourable tax rates. The entry of China into the WTO resulted in a trade boom as the country then had a wider access to global markets.

45 There are countries suffering from ongoing and widespread conflicts, which include civil war and ethnic violence (e.g. Darfur, Sudan). Many governments are corrupt and aid money does not reach the places for which it was intended. Dictatorships are common, which makes it difficult to resolve many issues, often leading to great instability. Several African countries are landlocked, which means that the movement of goods is often difficult, involving vast distances overland.

46 The EU began life in 1957 as the European Economic Community (EEC, also known as the 'Common Market') on the signing of the Treaty of Rome. The original six members were France, Italy, West Germany, Belgium, the Netherlands and Luxembourg. Its expansion then occurred at irregular intervals. The UK, Ireland and Denmark joined in 1973; Greece in 1981; Spain and Portugal in 1986; Finland, Sweden and Austria in 1995. The biggest expansion came in 2004 with 10 countries joining — Cyprus, Czech Republic, Estonia, Hungary, Latvia, Lithuania, Malta, Poland, Slovakia and Slovenia. Bulgaria and Romania became members in 2007.

47 The continued process of deregulation and liberalisation has allowed FDI into areas that were once dominated by state or domestic private-sector firms. A growing number, therefore, of the world's largest TNCs are to be found in the service sector, particularly banking, insurance and retailing organisations. The ICT revolution has also opened up overseas investment in tradeable services. IT-enabled services, for example, are now increasingly globalised because information can be sent easily around the world and service components can be located in the most efficient and cost-effective places. In many countries, therefore, the largest share of inward investment is now accounted for by the service sector.

48 When making the TNC case study, cover the history of the company, its characteristics (what it does, products etc.), spatial organisation and its impact both in its country of origin and where it now operates.

49 Try to find other NGOs that are involved in supplying aid to poorer countries (apart from Oxfam, which is given in the text).

50 Economic sustainability is the ability of economies to sustain themselves when resources decline (or become too expensive to exploit) and when the population dependent upon those resources is increasing. Environmental sustainability seeks to maintain ecological processes and life-support systems, preserve genetic diversity and ensure the utilisation of species and ecosystems without destroying them.

51 A conflict resulting from nationalism would be the product of extreme loyalty and devotion to a sovereign country. Examples would include Scottish and Welsh nationalism, or at the more violent extreme, that concerning Serbia after the break-up of the former Yugoslavia. Regionalism refers to consciousness of a homogeneous area within one or more countries. Examples include the Basques of Spain and France, and the Kurds of Turkey, Iraq, Iran and Syria.

52 Terrorism refers to the systematic use of terror as a means of coercion to a political, or more frequently an ideological, end. The 9/11 (2001) attack on the Twin Towers in New York was an act of terrorism. Insurrection is an act or instance of revolt against civil authority or an established government, usually involving rebellion against the rules of that government. Insurrection took place in a number of Arab nations during the spring of 2011.

53 Planning processes are more dominant in the UK. The Localism Bill, which passed through parliament in 2011, will mean that planning will originate from the local, rather than the national, level. Regional Development Authorities (RDAs) are being abolished and replaced with Local Enterprise Partnerships (LEPs) between local businesses and civic leaders. Regional strategies will be replaced by neighbourhood plans. The act also advocates a general presumption in favour of sustainable development.

54 An international conflict is one that involves more than one country, either as direct participants or as a provider of resolution, or as a receiver of refugees from the conflict.

55 A multicultural society is a social grouping that contains members from a wide variety of national, linguistic, religious or cultural backgrounds. It is often an emotive issue, especially when 'cultural' differences are interpreted as racial differences. Although skin colour remains as a visible distinguishing feature, people differ from one another in terms of ethnic differences, language, religion and culture.

56 One significant benefit that can be attributed to the multicultural society in the UK is the wide range of food types and outlets within the country. Most high streets now have Italian, Chinese and Indian restaurants, added to more recently by Mexican, Thai and even Polish. This is typified by the rise of the dish chicken tikka masala. The invention and origin of this dish has been attributed by some to a chef in Glasgow. Attempts have been made to have the European Union grant the dish Protected Designation of Origin status as a Glaswegian dish. Surveys have found chicken tikka masala to be the most popular meal in British restaurants and it has been called 'Britain's true national dish'.

57 Autonomy is the right of self-government. Separatism refers to the attempts by regional groups to gain more political control from central government over the area in which they live. For some groups, the ideal would be total independence.

58 Poverty can be measured by the international poverty line. This is based on a level of consumption representative of the poverty found in low-income countries. In 2008, the international poverty line was set at $1.25 a day, measured in terms of 1993 purchasing power parity (PPP). Poverty can also be measured by the Human Development Index (HDI) which combines the variables of life expectancy, educational attainment and real GDP per capita.

59 The G8 countries are the richest countries in the world: Japan, USA, Germany, France, Canada, UK, Italy and Russia.

60 An NGO is a non-governmental organisation which has not been created by a statutory Act but appears to have major responsibilities. NGOs are created mostly by volunteers, but are later supported by individual or corporate benefactors. Many are charities, and have international responsibilities such as organising relief and humanitarian operations, as in the case of the Red Cross. Greenpeace is an example of a NGO with significant worldwide following and influence.

61 Appropriate (intermediate) technology involves the matching of technology in the developing world to the needs and skills of the people of the area. Too many projects, supported by the developed world, and applied to the developing world, fail because the gap between the people's knowledge and the modern technology is too great. With a lower level of technology it is possible for people to be able to be taught the technical understanding and skills that they require to become self-sufficient.

62 The term failed state is often used by political commentators to describe a state perceived as having failed at some of the basic conditions and responsibilities of a sovereign government. The following attributes are often used to characterise a failed state:

- loss of physical control of its territory
- erosion of legitimate authority to make collective decisions
- an inability to provide reasonable public services
- an inability to interact with other states as a full member of the international community

A

aid 59–60, 71, 103–06
air masses 22, 24, 25
altitude 21
anticyclones 23, 25
asthenosphere 9
atmosphere 19–21

B

Bangladesh, global warming 28
biodiversity 38
  and development 89–91
  equatorial rainforest 34, 90–91
  tropical monsoon forest 36
  tropical savanna 35
biomes 30
  temperate 32–33
  tropical 33–36, 86–91
bottom-up projects 71
Brandt report 50
British Isles (*see also* UK)
  climate 21–23
  global warming 28

C

carbon emissions 29
car ownership and use 45, 48
central business district (CBD) 46–47
challenge, definition 62
children, health 70–71
China, economic growth 53
cities (*see also* inner city; urban areas)
  CBD 46–47
  environmental impact 94–96
  growth 93–94
  types 40, 92–93
City Challenge Partnerships 44–45
climate 19–30
  cool temperate western maritime 21–23, 32
  ecological responses to 88–89
  factors controlling 19–21
  tropical 24–25, 33, 34, 35, 83
  urban 26, 84–85
climate change 26–29
climatic climax 32
cold fronts 22–23
commuting 48
conflict 62–65, 108–10
conservation 38, 39
conservative margins 11, 12
constructive (divergent) margins 10, 12, 13
continental drift 9
cool temperate western maritime climate 21–23
Copenhagen Accord 29
core (Earth) 8
counter-urbanisation 41
crust (Earth) 8, 9, 10
culture 63, 108–09

D

debt, international 54
de-industrialisation 43, 52
depressions 22–23
destructive (convergent) margins 10–11, 12, 13
developing world (*see also* least developed countries; newly industrialising countries)
  poverty 69–71
development
  aid vs trade for 59–60, 71, 103–06
  and biodiversity 89–91
  continuum 50
  fragile ecosystems 38–39
  measurement 50, 69
  processes 52–53
  and security 72
  and sustainability 38, 60–61, 89–91
  TNC role 52–53, 102–03
  types 49–50
development gap 50–51

E

Earth, structure 8–9
earthquakes 11, 14–16, 17–19, 75–80
Earth Summit (Rio) 29, 60–61
economic development 49, 50
economic sustainability 60–61
ecosystems (*see also* biomes) 30
  British Isles 32–33
  development 32
  fragile 38–39
  human impact on 36–39
  structure 30
  trophic levels 31
  tropical 33–36, 86–91
  urban 36–38
ecotourism 61
emissions reduction 29
employment
  immigrants 67
  inner city 43
energy flows (ecosystem) 31
energy recovery 47
environment
  conflict affecting 109–10
  conservation 38, 39
  tourism impact 61
  urban impact 94–96
environmental lapse rate 21
environmental sustainability 60
equatorial climate 24, 83
equatorial rainforest 33–34, 86–91
ethnicity 63
ethnic minorities 66–67, 107–08
ethnic segregation 43, 66
European Union 55–56

F

fold mountains 11
food chains and webs 31

G

GDP 50
gentrification 42

Gersmehl diagrams 31
global groupings 54–56
globalisation (*see also* TNCs) 51–52, 100–101
global poverty 69–71
global shift 52
global warming 27–29
GNP 50
Goma 6
greenhouse gases 27
Gujerat earthquake 17–18

H
hazard management 18–19, 76–80
health, women and children 70–71
Heavily Indebted Poor Countries Programme 54
high-pressure weather systems 23
hot spots 11
housing 42, 43, 44, 66
human development index 50
hurricanes 25, 81–82

I
identity 62–63, 108–09
ideology 63, 108–09
immigration 65–67
India, economic growth 53
industrialisation, NICs 52–53
inequality (*see also* poverty) 43, 50–51
inner city
  decline 43–44
  regeneration 42–43, 44–45, 46–47
insurrection 63
international conflicts 65, 109–10
international debt 54
international groupings 54–56
international poverty line 69, 70
inter-tropical convergence zone (ICTZ) 25
island arcs 11
isobars 22, 23
ITCZ 25

K
Kyoto Protocol 29

L
landfill 48
land use conflicts 65
latitude 21
lava 13
lava flows 17
least developed countries (LDCs) 53–54, 59, 104–06
lithosphere 8, 9
Los Angeles earthquake 17–18
low-pressure weather systems 22–23

M
magma 9, 11
mantle 8
manufacturing 52, 56–57, 100–101
marketing, global 52
markets, emerging 53
Mercalli scale 15
mid-oceanic ridges 10
migration 40, 41, 44, 65–67
Millennium Development Goals 70–71
monsoon climate 24–25, 35
monsoon forest 35–36
Mount Etna 17
multicultural societies 65–67

N
natural hazards, management 18–19, 76–80
newly industrialising countries (NICs) 52–53
Northridge earthquake 17–18
North–South divide 50–51
nutrient cycling 31, 86–87
Nyiragongo eruption 16–17

O
ocean currents 21
ocean trenches 10, 11
overseas development aid 59–60, 71, 103–06

P
palaeomagnetism 10
plagioclimax 33
planning (urban) 65, 94–96
plant succession 32, 33, 37
plate margins 10–12, 13, 15
plate tectonics (*see also* earthquakes; volcanoes) 8–12
politics 63
population growth 40
poverty (*see also* inequality) 53–54, 69–71
precipitation 22, 23, 24
priseres 32
property-led regeneration 42

R
rainforest, equatorial 33–34, 86–91
recycling 29, 47
resources, conflict over 64–65
retailing 45–46, 58–59
re-urbanisation 42–43
Richter scale 15
rift valleys 10
Rio Earth Summit 29, 60–61
road traffic 48–49
rural areas, migration to 41
rural–urban fringes 37–38
rural–urban migration 40

S
savanna 24, 34–35
sea-floor spreading 10
secondary succession 33, 37
security, and development 72
segregation 43, 66
seismicity *see* earthquakes

separatism 67–69, 111–13
seres 3
services, globalisation 52, 57, 58–59
shopping 45–46
soils
  temperate 32–33
  tropical 34, 35, 36
storms
  mid-latitude 23
  tropical 25, 81–83
suburbanisation 41
succession, plant 32, 33, 37
sustainability
  and development 38, 60–61, 89–91
  ecosystem management 38–39
  urban areas 42, 43, 47–48
sustainable tourism 61

T

tectonic activity *see* plate tectonics
temperate deciduous woodland 32–33
temperature 21
  cool temperate climate 22
  trends 27
  tropical climates 24
  urban climates 26
territory 63
terrorism 63
Tesco 58–59
TNCs 52, 56–59, 100–103
top-down projects 71
tourism, sustainable 61
Toyota 100–101
trade 52, 59–60, 71, 103–06
trade blocs 54–55
traffic management 48–49
transnational corporations (TNCs) 52, 56–59, 100–103
tri-cellular model 20
trophic levels 31
tropical biomes 33–36, 86–91
tropical monsoon climate 24–25
tropical monsoon forest 35–36
tropical rainforest 33–34, 86–91
tropical revolving storms 25, 81–83
tropical savanna grasslands 34–35
tropical wet/dry savanna climate 24
tsunamis 15–16

U

UK
  City Challenge Partnerships 44–45
  climate 21–23
  emissions reduction 29
  ethnic minorities 107–08
  global warming 28
  land use conflicts 65
  multicultural societies 65–67
  TNCs 58–59
  urban decline and regeneration 42–44, 97–99
  waste management 47–48
United Nations
  Human Development Report 69
  Millennium Development Goals 70–71
  Rio Earth Summit 29, 60–61
urban areas
  climate 26, 84–85
  decline 43–44
  ecosystems 36–38
  environmental impact 94–96
  growth 93–94
  migration from 41, 44
  plant succession 37
  regeneration 42–43, 44–45, 46–47, 96–99
  retailing 45–46
  sustainability 42, 43, 47–49
  transport management 48–49
  waste management 47–48
Urban Development Corporations (UDCs) 42, 97
urbanisation 39–41
urban niches 37

V

vegetation
  climatic climax 32
  equatorial rainforest 33–34, 88–89
  plagioclimax 33
  succession 32, 33, 37
  temperate deciduous woodland 32
  tropical monsoon forest 35–36
  tropical savanna 34–35
  urban 36–38
volcanic landforms 12–14
volcanoes
  distribution 10, 11, 12
  impact 14, 16–17, 78–80
  prediction and management 18

W

war 63, 110
warm fronts 22–23
waste management 47–48
weather systems (*see also* climate) 22–23
Wegener, Alfred 9
winds 20, 22